# WHY THE
# GREEN NEW DEAL
# IS A BAD DEAL FOR AMERICA!

## By Martin Capages Jr. PhD

With Foreword by Andy May
Author of Climate Catastrophe! Science or Science Fiction?

© 2019 Martin Capages Jr

American Freedom Publications LLC

www.americanfreedompublications.com

2638 E. Wildwood Road

Springfield 65804

ISBN 978-1-64516-432-6 Paperback Version

ISBN 978-1-64516-433-3 eBook Version

Cover Design     Christopher. M. Capages

www.capagescreative.com

First Edition- February 18, 2019

# DEDICATION

## TO THE ENERGY PROVIDERS OF AMERICA

# OTHER WORKS BY THE AUTHOR

BOOTS TO BOGIES TO BRONZE: The Authorized World War II Biography of 2LT Jack C. Pyatt

THE MORAL CASE FOR AMERICAN FREEDOM

OZARK COUNTY HEART: Boyhood Memories of a Dora Missouri Farm

A WAKEFUL WATCH: The Authorized Biography of Charles Lindbergh Armstrong

HEARTLAND REBELLION

THE SILENT SECOND: The Biography of Martin Capages- Captain USMC

EPIPHANY: Before Time Zero- Faith of an Engineer

# ACKNOWLEDGEMENTS

I would like to thank the following people who contributed so much time and effort during the preparation of this book:

To my wife Pamela for taking time from her own devotions and writings to offer her encouragement, thoughts, patience and prayers during the preparation of this, yet another, book by her husband,

To Mr. Andy May, Author of *CLIMATE CASTASTROPHE!: Science or Science Fiction?*, for his technical input and for his willingness to write the Foreword to this book,

To Dr. Ronald R. Cherry, Author of *RESTORING THE AMERICAN MIND*, for his continued support, friendship and leadership in all matters of Faith and Patriotism,

To the Federal employees of the United States Energy Information Administration for the outstanding work they do in providing important material and data on Energy to the American public.

# Table of Contents

# FOREWORD

Dr. Martin Capages has done an excellent job of deconstructing the "Green New Deal." Alexandria Ocasio-Cortez is only the latest, in a long line of politicians, to use climate change as an excuse for world government and global control of production, distribution and exchange of goods and services, aka socialism. The global warming (or climate change, if you prefer) scare has been inexorably tied to socialism since it was conceived in the late 1980s by Maurice Strong. He became the founding director of the UN Environmental Programme (UNEP) and later, in 1992, he created the UN Framework Convention on Climate Change. The UN Framework Convention on Climate Change has dominated (some would say "dictated") the global climate change agenda ever since.

But, the global warming/climate change is not a scientific issue, it is an economic and political one. By speculating that climate change is man-made, through our carbon dioxide emissions, and dangerous, the politicians can claim that to save the planet we must form a global governmental body to reduce carbon dioxide emissions and "save the planet."

However, whether climate change is mostly natural or mostly man-made, is less important than the rate of the change and if it is dangerous. The rate of global warming over the past 150 years is less than one degree Celsius (1.8°F) per 100 years, this is not alarming and, if anything, it appears to be slowing down in recent decades (Fyfe, et al., 2016). So, a climate catastrophe is not headed our way anytime in the next few hundred years, we do have plenty of time to study the matter.

Characterizing the current warming as an urgent and impending crisis is silly considering the scientific evidence we have today. There is no need to remove national boundaries, form a global government and abandon capitalism to "save the world." Climate changes, we all accept this, perhaps it is mostly man-made, perhaps it is mostly natural, we don't know. What we do know is that many communities may be affected by climate change. Sea level is rising, the best long-term estimates are that it is rising between 1.8 and 3 millimeters per year. This is not a large rate, perhaps seven inches to a foot in 100 years, much less than the daily tides. But, if it causes problems, seawalls can be built, people can move from dangerous areas or elevate their houses, it is a problem that can be dealt with locally, as it has been for thousands of years.

With fossil fuels or nuclear power, which the climate change alarmists want to eliminate, we can cool or heat our buildings if a community gets too cold or too hot. If we get more rain, we can improve our drainage or move out of flood plains. If it gets too dry, we can drill wells for water or move water via aqueducts. The point is, each community needs to deal with its own problems. Climate change is not a problem that must be dealt with globally, the people affected and closest to the problem will deal with it in the most effective and efficient manner, as they always have.

So, consider Dr. Capages arguments carefully. Capitalism built our current affluent society and lifted billions of people out of abject poverty. Do we really need to throw all of this away and turn all our businesses and property rights over to a world government to fight a possible climate change problem that is hundreds of years away, if it exists at all?

---Andy May
      Author of *Climate Catastrophe! Science or Science Fiction?*

# PREFACE

The Green New Deal utilizes a propaganda technique called Appeal to Fear. It advances the notion that the U. S. must initiate a massive *and urgent* State-run program that would control the climate by reducing the use of carbon-based fuels. It would simultaneously improve the lot of the poor and middle class. In truth, this proposed program would be a complete waste of human and natural resources. The entire concept is based on the false premises that carbon dioxide ($CO_2$) is a pollutant, $CO_2$ emissions are bad, that the increasing rate of $CO_2$ emissions from burning fossil fuels is causing the global temperature to rise which will lead to adverse climate change, and that reducing fossil fuel use is an immediate necessity to protect the natural environment from most human actions. None of the premises are true but an entire political party has been co-opted by the 'old' glamour of young, naïve want-to-be Marxists who have seized on this invalid concept, The New Green Deal, to disguise a push for complete State control in all matters. The Party establishment has caved and now Democratic Socialism is a primary plank in the Party's progressive political platform.

This book is intended to address the future role of carbon-based fuels in a rational manner and without a hidden political or ideological agenda. It is agreed that the supply of fossil fuels is finite and the natural environment is affected by its continued use. Most of the effects are beneficial for both humans and the environment. But some are not. However, there is enough time to plan for a transition to cleaner energy without massive government intervention in the form of the control of production and the private sector. That type of government intervention is by definition-- Socialism. It has never been successful.

It is unfortunate that today labels such as "climate denier" or "climate contrarian" are being used to great effect against those asking some reasonable questions about the state-of-the-art of climatology, the accuracy of climate models, the scientific conclusions and the recommended actions derived from those conclusions. To challenge a consensus is what science is all about. I'm an engineer, not a scientist. But that does not prevent me from asking logical questions about the science and why, in particular, there is such a rush to transition away from the use of carbon-based fuels. That does not make me or anyone else of the same mindset "deniers or contrarians".

This is a serious matter. One that must be considered in a clear-eyed and non-ideological way. This is not a liberal arts matter, it must be considered pragmatically, with constructive thought and action. And the first practical thought is to recognize that we have plenty of time to gather the data and get it right, the American Way, not the Democratic Socialists Green New Deal way.


---Martin Capages Jr. PhD PE

Author of *The Moral Case for American Freedom*

# INTRODUCTION

Humankind has survived and flourished in a world beset with hazards. Its survival has been aided by its superior intelligence that allowed humankind to leverage the force multiplier of accessible energy. The sun has always provided the energy base but its areal peculiarities and timing variances have created poor living conditions in most locations. Survival has required the successful harnessing of energy from a source other than direct sunlight. The primary alternative source of energy came from the burning of wood, peat, dung and coal. Other energy sources came from the capture of wind and water flow. This allowed the population to grow.

That was the way it was until the beginning of the Industrial Age. Then humankind learned how to harness the energy of combustion by producing steam, first to drive pistons, then turbines. The initial primary fuel (after wood, dung and peat) was coal, then fuel oil and manufactured gas were added to the mix of carbon-based fuels. In the early stages, the primary benefit was mechanical advantage in production processes. But eventually, the primary end product became electricity, what some call secondary energy. Electricity became the primary driver of technological progress. It provided freedom from manual labor which could then be directed to entrepreneurial thought. This led to scientific research that yielded improvements in living conditions.

Electricity could be produced with mechanical generators rotated by hydropower and steam power. The steam was originally produced by burning fossil fuels, primarily coal. Scientific research then developed nuclear power to provide the steam while new turbines were designed to be powered by natural gas. The increased availability of natural gas due to breakthroughs in drilling and production methods along with its lower environmental effects caused a rapid reduction in the use of coal for electrical generation. Nuclear power growth was stagnated by fear and ignorance. Misguided government policies on powerplant emissions accelerated the coal-fired plant declines; however, improvements in emission controls, recently revised government policies and increased world demand have slowed the rate of decline somewhat.

The abundant available energy allowed the development and manufacturing of photoelectric cells arranged in banks of solar panels that can now use the available sunlight to effectively produce electricity in some locations. Both international and domestic engineering improvements in wind turbine design have also improved the viability of this older technology. However, small scale solar and wind generated power today are not economically viable in most locations without government subsidies. In addition, both wind turbines and solar panels provide intermittent, even unreliable power that requires either a means to store energy or a backup power source. The backup source is usually a fossil-fueled or nuclear power generation plant. The most reliable back-up power source is a natural gas-fired, gas turbine plant due to its ability to respond quickly. Coal-fired and nuclear plants are less flexible or efficient.

Those are the facts. But facts do not seem to matter when political ideologies enter the mix of energy source discussions. It should be understood though that political ideologies did not initiate the Industrial Age, common sense and entrepreneurialism were the driving forces. The goal was to survive and improve one's lot in life. The economic process that supported the entrepreneurial effort required was, and still is, Capitalism.

Today, in the era of commercialized 24/7 mainstream media reporting, common sense does not increase viewer ratings. A common-sense advocate will be labeled as a pejorative '____ denier.' But the thing about common sense is that it will eventually be recognized as Truth. Unfortunately, that will not reduce the enormous monetary losses incurred and human suffering that will result by following the wrong path led by the latest "shiny thing." That shiny thing is Democratic Socialism dressed up as The Green New Deal. Now a goal is being set to get to 100 percent Renewable Energy as soon as possible. But why? And what does renewable mean? And shouldn't it be reliable too?

Renewable seems to imply that the energy used is being replenished, fully replenished, after each use. How long is that replenishment process supposed to take? Being reliable is important. Green energy is not reliable. It must always be produced in surplus so that some of it can be stored in some manner or it must have an alternative to back it up when the sun doesn't shine or the wind doesn't blow. Again, these are just facts.

Recent discoveries and process improvements have made fossil fuels (petroleum and natural gas) more abundant than ever and less expensive than any other energy alternative. In combination with coal, this energy source, fossil fuels, will be available for centuries. We have time to move to a better energy source if we use common sense. This means restricting or even eliminating sources of pollution that impact the ability of humans to flourish in a challenging natural world environment. In the process, we need to be sure that we identify pollutants accurately without surrendering the Scientific Method to ideologies promoted by political activists with hidden agendas.

In my opinion, the eventual ultimate energy source could turn out to be nuclear fusion with both electricity and hydrogen as the secondary energy products. Electricity, as the energy medium, would also be a hydrogen generator while hydrogen would be an electricity generator. Again, that's my opinion. It will take major capital investments in research to make nuclear fusion and hydrogen fuel cells the end game.

We should examine the proposed Green New Deal legislation but not dismiss it without a counterproposal that would achieve the right objective, a logical transition to clean energy without tearing down the founding principle of this great country, the right to Individual Freedom protected by a constitutionally Limited Government.

# THE GREEN NEW DEAL

The Green New Deal is a proposed economic stimulus program in the United States that aims to address both economic inequality and climate change. The name refers to the New Deal, a combination of social and economic reforms and public works projects undertaken by President Franklin D. Roosevelt in response to the Great Depression. Supporters of a Green New Deal advocate a combination of Roosevelt's economic approach with modern ideas such as renewable energy and resource efficiency. (Wikipedia, 2019)

The Green New Deal that freshman U. S. Congressional Representative from the 14th District of New York, 29 year-old Alexandria Ocasio-Cortez has laid out aspires to power the U.S. economy with 100 percent renewable energy within 12 years and calls for "a job guarantee program to assure a living wage job to every person who wants one," "basic income programs" and "universal health care," financed, at least in part, by higher taxes on the wealthy.

## PROPOSED LEGISLATION

A draft of the proposed legislative bill has been widely circulated on the Internet by the drafters and supporters. The bill, as drafted, has the following scope:

(A) The Plan for a Green New Deal (and the draft legislation) shall be developed with the objective of reaching the following outcomes within the target window of 10 years from the start of execution of the Plan:

Dramatically expand existing renewable power sources and deploy new production capacity with the goal of meeting 100% of national power demand through renewable sources; building a national, energy-efficient, "smart" grid; upgrading every residential and industrial building for state-of-the-art energy efficiency, comfort and safety; eliminating greenhouse gas emissions from the manufacturing, agricultural and other industries, including by investing in local-scale agriculture in communities across the country; eliminating greenhouse gas emissions from, repairing and improving transportation and other infrastructure, and upgrading water infrastructure to ensure universal access to clean water; funding massive investment in the drawdown of greenhouse gases; making "green" technology, industry, expertise, products and services a major export of the United States, with the aim of becoming the undisputed international leader in helping other countries transition to completely greenhouse gas neutral economies and bringing about a global Green New Deal.

(B) The Plan for a Green New Deal (and the draft legislation) shall recognize that a national, industrial, economic mobilization of this scope and scale is a historic opportunity to virtually eliminate poverty in the United States and to make prosperity, wealth and economic security available to everyone participating in the transformation. In furtherance of the foregoing, the Plan (and the draft legislation) shall:

(i.) provide all members of our society, across all regions and all communities, the opportunity, training and education to be a full and equal participant in the transition, including through a job guarantee program to assure a living wage job to every person who wants one;

(ii.) diversify local and regional economies, with a particular focus on communities where the fossil fuel industry holds significant control over the labor market, to ensure workers have the necessary tools, opportunities, and economic assistance to succeed during the energy transition;

(iii.) require strong enforcement of labor, workplace safety, and wage standards that recognize the rights of workers to organize and unionize free of coercion, intimidation, and harassment, and creation of meaningful, quality, career employment;

(iv.) ensure a 'just transition' for all workers, low-income communities, communities of color, indigenous communities, rural and urban communities and the front-line communities most affected by climate change, pollution and other environmental harm including by ensuring that local implementation of the transition is led from the community level and by prioritizing solutions that end the harms faced by front-line communities from climate change and environmental pollution;

(v.) protect and enforce sovereign rights and land rights of tribal nations;

(vi.) mitigate deeply entrenched racial, regional and gender-based inequalities in income and wealth (including, without limitation, ensuring that federal and other investment will be equitably distributed to historically impoverished, low income, deindustrialized or other marginalized communities in such a way that builds wealth and ownership at the community level);

(vii.) include additional measures such as basic income programs, universal health care programs and any others as the select committee may deem appropriate to promote economic security, labor market flexibility and entrepreneurism; and

(viii.) deeply involve national and local labor unions to take a leadership role in the process of job training and worker deployment.

(C) The Plan for a Green New Deal (and the draft legislation) shall recognize that innovative public and other financing structures are a crucial component in achieving and furthering the goals and guidelines relating to social, economic, racial, regional and gender-based justice and equality and cooperative and public ownership set forth in paragraphs (2)(A)(i) and (6)(B). The Plan (and the draft legislation) shall, accordingly, ensure that the majority of financing of the Plan shall be accomplished by the federal government, using a combination of the Federal Reserve, a new public bank or system of regional and specialized public banks, public venture funds and such other vehicles or structures that the select committee deems appropriate, in order to ensure that interest and other investment returns generated from public investments made in connection with the Plan will be returned to the treasury, reduce taxpayer burden and allow for more investment.

The drafters of this plan have anticipated the questions that may be asked about the proposed legislation and have developed some answers as follows:

| FREQUENTLY ASKED QUESTIONS |
| --- |
| |
| *Why do we need a sweeping Green New Deal investment program? Why can't we just rely on regulations and taxes alone, such as a carbon tax or an eventual ban on fossil fuels?* |
| • Regulations and taxes can, indeed, change some behavior. It's certainly possible to argue that, if we had put in place targeted regulations and progressively increasing carbon and similar taxes several decades ago, the economy could have transformed itself by now. But whether or not that is true, we did not do that, and now time has run out. |
| • Given the magnitude of the current challenge, the tools of regulation and taxation, used in isolation, will not be enough to quickly and smoothly accomplish the transformation that we need to see. |
| • Simply put, we don't need to just stop doing some things we are doing (like using fossil fuels for energy needs); we also need to start doing new things (like overhauling whole industries or retrofitting all |

buildings to be energy efficient). Starting to do new things requires some upfront investment. In the same way that a company that is trying to change how it does business may need to make big upfront capital investments today in order to reap future benefits (for e.g., building a new factory to increase production or buying new hardware and software to totally modernize its IT system), a country that is trying to change how its economy works will need to make big investments today to jump-start and develop new projects and sectors to power the new economy.

• The draft resolution sets out a (non-exhaustive) list of several major projects that need to be completed fast. These include upgrading virtually every home and building for energy efficiency, building a 100% greenhouse gas neutral power generation system, decarbonizing industry and agriculture and more. These projects will all require investment. • We're not saying that there is no place for regulation and taxes (and these will continue to be important tools); we're saying we need to add some new tools to the toolkit.

*Why should the government have a big role in driving and making any required investments? Why not just incentivize the private sector to invest through, for e.g., tax subsidies and such?*

• Two main reasons: (1) scale and (2) time. **First - scale**. The level of investment required will be massive. Even if all the billionaires and companies came together and were willing to pour all the resources at their disposal into this investment, the aggregate value of the investments they could make would not be sufficient. For example, the "$1 trillion over 10 years" plan for investment in the green economy that has been floated by some policy makers has been criticized by climate experts as a wholly inadequate level of investment - $1 trillion is the entire market cap of Amazon, one of the biggest companies of all time (and it is far ahead of its closest competitors in terms of market size). **Second - time.** The speed of investment required will be massive. Even if all the billionaires and companies could make the investments required, they would not be able to pull together a coordinated response in the narrow window of time required to jump-start major new projects and major new sectors. Time-horizons matter in another way - by their nature, private companies are wary of making massive investments in unproven research and technologies; the government, however, has the time horizon to be able to patiently make investments in new tech and R&D, without necessarily having a commercial outcome or application in mind at the time the investment is made. Major examples of government investments in "new" tech that subsequently spurred a boom in the private section include DARPA-projects, the creation of the internet - and, perhaps most recently, the government's investment in Tesla.

- We've also seen that merely incentivizing the private sector doesn't work - e.g. the tax incentives and subsidies given to wind and solar projects have been a valuable spur to growth in the US renewables industry but, even with such investment-promotion subsidies, the present level of such projects is simply inadequate to transition to a fully greenhouse gas neutral economy as quickly as needed.

- Once again, we're not saying that there isn't a role for private sector investments; we're just saying that the level of investment required will need every actor to pitch in and that the government is best placed to be the prime driver.

*How will the government pay for these investments?*

- Many will say, "Massive government investment! How in the world can we pay for this?" The answer is: in the same ways that we paid for the 2008 bank bailout and extended quantitative easing programs, the same ways we paid for World War II and many other wars. The Federal Reserve can extend credit to power these projects and investments, new public banks can be created (as in WWII) to extend credit and a combination of various taxation tools (including taxes on carbon and other emissions and progressive wealth taxes) can be employed.

• In addition to traditional debt tools, there is also a space for the government to take an equity role in projects, as several government and government-affiliated institutions already do.

*Why do we need a select committee? We already have committees with jurisdiction over the subject matter e.g. Energy and Commerce, Natural Resources and Science, Space and Technology. Just creating another committee seems unnecessary.*

• This is a big problem with lots of parts to it. The very fact that multiple committees have jurisdiction over parts of the problem means that it's hard for any one of those existing committees to generate a comprehensive and coherent plan that will actually work to transform America's economy to become greenhouse gas neutral in the time we have left.

• Not having a full 360° view of, and approach to, the issue (and only having authority over a part of the issue) means that standing committee solutions would be piecemeal, given the size and scope of the problem. A Democratic administration and Congress in 2020 will not have the time to sort through and combine all those solutions in the brief window of opportunity they will have to act.

• Select committees, in the Congressional Research Services' own words, serve the specific function of "examining emerging issues that do not fit clearly within existing standing committee jurisdictions or cut across jurisdictional boundaries."

• The challenges that the Select Committee For A Green New Deal is mandated to meet fit squarely within this space.

• This does not need to be a zero-sum proposition between committees. Just as Markey-Waxman was collaborative between the head of the Select Committee and standing Energy & Commerce committee, this can also be collaborative. More is more. A select committee ensures constant focus on climate change as the standing committee deals with that and many other issues of the day -- such as wild fires in California, Infrastructure, clean water issues, etc.

*Why should we not be satisfied with the same approach the previous select committee used (i.e. the Select Committee on Energy Independence and Global Warming)? Why do we need a new approach?*

• The previous select committee did not have a mandate to develop a plan for the transformation of our economy to become carbon neutral. It mainly held hearings to draw attention to the problem of climate change. That was already too little too late in 2007-11 when the committee was active.

• The previous select committee's work can be summarized as follows: The "sole authority" it did have was to "investigate, study, make findings, and develop recommendations on policies, strategies, technologies and other innovations, intended to reduce the dependence of the United States on foreign sources of energy and achieve substantial and permanent reductions in emissions and other activities that contribute to climate change and global warming." From March 2007 to December 2010 - a full 3.5 years - they did the job that they were tasked to do and held hearings and prepared reports. Per their website, they "engage[d] in oversight and educational activities through hearings, reports, briefings and other means intended to highlight the importance of adopting policies which reduce our dependence on foreign oil and our emissions of global warming pollution."

- So there has already been a select committee that did the investigating to highlight that it was important to have some action on this issue - it's now time to move on from investigating and reporting to action.

- The old select committee also had (even within its limited investigative mandate) the limitation that it focused on strategies for reducing foreign energy dependence and reducing emissions - rather than treating climate issues as the integrated social, economic, scientific challenge that it is.

*Why does this new select committee need to prepare draft legislation? Isn't investigation, hearings, briefings and reporting enough?*

- The old select committee was mandated merely to investigate and prepare reports for other people and House Committees to read and act on.

- The idea was that (as per the old select committee's website) "each Member of the Select Committee sits on legislative committees which process legislation and amendments affecting energy independence and global warming issues in other committees" and presumably, that those members would take the work of the select committee and come up with legislation in their own committees.

- However, this approach did not make a big impact relative to the scale of the problem we face. The one piece of legislation that eventually came out of the old select committees work - the American Clean Energy and Security Act of 2009 (ACES) https://www.markey.senate.gov/GlobalWarming/legislation/index.ht

13

ml) was a cap-and-trade bill that was wholly insufficient for the scale of the problem.

• The House had a chance (from 2007 to 2010) to try a version of a select committee that investigated an issue and then passed along preparation of legislation to other committees - the result of that process doesn't inspire any confidence that the same process should be followed again if we wish to draft a plan to tackle the scale of the problem we face.

• The new select committee will also continue to have investigative jurisdiction, so the new proposal isn't taking anything away from the old one - it is adding things on to make the select committee more effective.

*What's an example of a select committee with abilities to prepare legislation? Does the new Select Committee For A Green New Deal seem to fit on that list?*

• Recent examples for select committees in the House include: Ad Hoc Select Committee on the Outer Continental Shelf (94th-95th Congresses), Ad Hoc Select Committee on Energy (95th Congress), Select Committee on Homeland Security (107th Congress), and Select Committee on Homeland Security (108th Congress).

• The Congressional Research Service notes (in discussing these four recent select committees with legislative jurisdiction) that "The principal explanation offered in creating each of the four select committees with legislative authority was that their creation solved jurisdictional problems. The proponents in each case indicated that multiple committees claimed jurisdiction over a subject and that the House would be unable to legislate, or at least to legislate efficiently, in the absence of a select committee."

• The proposed subject matter and mandate for the Select Committee For A Green New Deal sits squarely within this general description for a select committee with the ability and mandate to prepare legislation.

*Doesn't this select committee take away jurisdictional power from the other (standing i.e. permanent) committees that have jurisdiction over at least part of the issue?*

• All of the relevant standing committees will be able to provide input to and make their wishes known to the select committee during the creation of both the plan as well as the draft legislation, and then in a future Congress, when it comes to crafting and passing the final legislation, that Congress can take a decision on the best mechanism for bringing that final legislation to a floor vote and passage.

• Allowing the select committee to draft legislation doesn't take any jurisdiction away from current standing committees, it is entirely additive.

• The legislation developed by the select committee would still need to be referred to and pass through the permanent House Committees that have jurisdiction over parts of the subject matter.

• For example, the legislation drafted by the Select Committee on Homeland Security needed to pass through the permanent committees on Agriculture; Appropriations; Armed Services; Energy and Commerce; Financial Services; Government Reform; Intelligence (Permanent Select); International Relations; Judiciary; Science; Transportation and Infrastructure; Ways and Means.

• The benefit of a select committee in this case would also be that there would be a single forum that could act as a quarterback in working through and resolving any comments or issues brought up by the other House Committees, which would streamline the process of drafting this legislation.

*But a select committee only exists for the congressional session that created it! So even if this select committee prepares legislation, it likely won't get passed in this session by a Republican-held Senate and White House, so why does having a select committee now even matter?*

• The proposed new select committee would work in two stages (which wouldn't necessarily have to be sequential): o First, they would put together the overall plan for a Green New Deal - they would have a year to get the plan together, with the plan to be completed by January 1, 2020. The plan itself could be in the form of a report or several reports. o Second, they would also put together the draft legislation that actually implements the plan - they could work on the draft legislation concurrently with the plan (after they get an initial outline of the plan going) and would need to complete the draft legislation within 90 days of completing the plan (i.e. by March 1, 2020 at the latest) o The select committee is also required to make the plan and the draft legislation publicly accessible within 30 days of completing each part

• The plan and the draft legislation won't be developed in secret - they are specifically required to be developed with wide and broad consultation and input and the select committee can share drafts or any portions of their work with the other House Committees at any time and from time to time, so their work will be conducted in the open, with lots of opportunities to give input along the way.

• The idea is that between (a) developing the plan and the draft legislation (and holding public hearings and briefings along the way as needed), (b) the plan coming out in Jan 2020 and (c) the draft legislation coming out in March 2020, the relevant permanent House Committees, House members, experts and public will have time to engage with, discuss, revise the draft legislation between March 2020 and the end of the 116th Congress so that, by the end of this congressional term, there

16

is a comprehensive plan and enacting legislation all lined up as soon as the new (Democratic) Congress convenes in January 2021.

*What's wrong with the other proposed legislation on climate change? Can't we just pass one of the other climate bills that have been introduced in the past? Why prepare a whole new one?*

• The shortest and most accurate response is that (1) none of them recognize the extent to which climate and other social and economic issues are deeply interrelated and (2) even if looking at climate as a stand-alone issue, none of them are scaled to the magnitude of the problem.

• Of the other proposed legislation, the OFF Act could be a good starting point. (AOC Green New Deal, 2019)

[OFF Act. Also known as the OFF Act, the bill would place a moratorium on new fossil fuel projects and moves our electricity and most transportation systems to 100% renewable energy by 2035. It would also provide for a truly just transition for environmental justice communities and those working in the fossil fuel industry. [According to its proponents] (Rep. Gabbard, Tulsi [D-HI-2], 2017)

# ANALYSIS OF PROPOSED DRAFT LEGISLATION

A review of the Frequently Asked Questions and the Answers is perhaps more revealing than the actual draft proposal. There is an unveiled disappointment and anger at the previous Democratic leadership, both when in the majority and the minority. The Democratic Socialists have formed a coalition to attack their perception of pervasive social injustice. They are using imagined catastrophic climate change as the primary reason for the legislation. They are exceedingly disappointed in past efforts by the established Democrat politicians, in particular the committee chairpersons and minority leaders. So, they propose a select committee of themselves to make the changes they feel are urgently needed. Their reasoning is "The previous select committee did not have a mandate to develop a plan for the transformation of our economy to become carbon neutral. It mainly held hearings to draw attention to the problem of climate change."

Let's begin with the "problem of climate change." There is no human action that can stop or significantly mitigate climate change. The real problem will be wasting human and monetary resources in a senseless attempt to stop climate change when those resources will be necessary to fight the actual impact of adverse climate change. And, we will most likely need to mitigate the effects of real climate change attributable to global cooling, not global warming. However, there is no urgency in either case.

But, the proposed legislation implies that it is an urgent matter … "if we had put in place targeted regulations and progressively increasing carbon and similar taxes several decades ago, the economy could have transformed itself by now. But whether or not that is true, we did not do that, and now time has run out." This also incorrect. There is plenty of time for a reasonable transition from fossil fuels to a cleaner energy source.

In 1980, the U. S. Reserve to Production ratio(R/P) was suggested to mean we only had 32 years of oil production from existing reserves. We are well past the year 2012 now. With technological advances, reserves that were previously deemed unviable to tap have kept coming on board. Various studies now show that the total worldwide remaining recoverable oil resources would last 190 years, natural gas 230 years, and coal, a whopping 2900 years. (Singh, 2015)

The Green New Deal asserts that this must be a Government program. The two main reasons are the huge scale of the program and resources required and the urgent timeline to achieve the result, 100% renewable energy in 10 to 12 years. The proponents believe that private industries' focus on their bottom lines will prevent the necessary investments in research required. They tout that the Government does not have that problem and will just "tax and print money [to achieve the goal.]

The Green New Deal proponents pull on "old Democrat" heartstrings to relive the glory of FDR's New Deal that they believe ended the Great Depression. But they are wrong. Two UCLA economists say they have figured out why the Great Depression dragged on for almost 15 years, and they blame a suspect previously thought to be beyond reproach: President Franklin D. Roosevelt. After scrutinizing Roosevelt's record for four years, Harold L. Cole and Lee E. Ohanian conclude in a new study that New Deal policies signed into law 71 years ago thwarted economic recovery for seven long years.

"Why the Great Depression lasted so long has always been a great mystery, and because we never really knew the reason, we have always worried whether we would have another 10- to 15-year economic slump," said Ohanian, vice chair of UCLA's Department of Economics. "We found that a relapse isn't likely unless lawmakers gum up a recovery with ill-conceived stimulus policies." (Sullivan, 2004)

The Democratic Socialists obviously do not know want to know the truth of their own history. They propose, "In addition to traditional debt tools, there is also a space for the Government to take an equity role in projects, as several Government and Government-affiliated institutions already do." This is also termed nationalizing (or fascism and socialism) but the drafters of this legislation probably know that. It's what this legislation is all about anyway.

The rest of the question answers are directed at the steps to get this legislation ready for a Democrat-controlled Congress and Executive Branch in 2020.

Now we should look at the scope of the proposed legislation in more detail. The goal is to "Dramatically expand existing renewable power sources and deploy new production capacity with the goal of meeting 100% of national power demand through renewable sources." This goal is not achievable. Renewables currently contribute 11 percent to total capacity. Biomass, though renewable (about 45% of total renewables), will increase $CO_2$ emissions (and black carbon emissions) since they require burning. Solar and wind are contributing less than 3% to total capacity (in 2017). The maximum contribution of wind and solar will not exceed 20% even if viable and scalable energy storage technologies become reality.

The next stated objective is "building a national, energy-efficient, 'smart' grid. This is a great objective. Might we add that it should also be EMP resistant as well as have a lot of nuclear, clean-coal and natural gas turbine power plant nodes attached.

Next on the list we have "upgrading every residential and industrial building for state-of-the-art energy efficiency, comfort and safety."

Hard to argue with this either, except, who is doing the upgrading? Who is paying for the upgrading? If it's me, then what if I don't want to? Another IRS-enforced mandate?

Next, we have "eliminating greenhouse gas emissions from the manufacturing, agricultural and other industries, including by investing in local-scale agriculture in communities across the country." The problem with this is that fertilizers including $CO_2$ itself, are necessary ingredients for local-scale agriculture, not to mention large-scale agriculture. Most of the fertilizer is from carbon-based liquids and gas, transported and even injected into the soil with fossil fuel-driven equipment. Harvesting equipment is also very fossil fuel-driven. This infrastructure is not going to change much in 10 years, or 30 for that matter. And here we have another conflict between those of anti-GMO and pro-Organic mindset and the unwashed masses who need to eat to survive. Productivity in agriculture is important to farmers.

As we continue with this legislative punch list, we have "eliminating greenhouse gas emissions from, repairing and improving transportation and other infrastructure, and upgrading water infrastructure to ensure universal access to clean water." While the "eliminating greenhouse gas emissions from transportation" is consistent with the primary goal, it is not clear where "improving other infrastructure including universal access to clean water" comes into play within the scope of the legislation.

The rest of the scope is essentially spending enough money to put lipstick on this socialistic pig and then try to market it to the world as the latest Made in America fad. I don't think China or India are going to buy into the idea. They need energy. And I doubt that Spain, France, Germany or even the UK are going to be fooled again. They've already been there themselves. They just don't want to talk about it.

But legislation is legislation, even if it is drafted by nitwits and supported by the liberal main stream media. Here are some examples:

A team of LEGO experts spent around 600 hours building the world's Largest LEGO® brick wind turbine and then assembling it outside the Liverpool ONE shopping center in Liverpool, UK. The Danish company attempted this challenge to celebrate the fact that LEGO now balances 100% of its energy use with energy from renewable sources, since investing in Liverpool's Burbo Bank Extension wind farm.

The famous toy company also wanted to show kids the importance of using renewable energy and learning about environmental issues. Now it is not clear what "balances 100% of its energy use with energy from renewable sources" actually means, we can guess that, perhaps, they have a carbon offset agreement with the UK national grid. Let's not waste time researching that possibility. We will just assume that the wind does not blow all the time at the Burbo Bank, off the west shores of the UK in the Irish Sea. I know, I have been there in a Force 8 and also in a dead calm. That was a long time before the wind farm was erected, but I assume not much has changed in the Irish Sea since then. The point is, there are times when energy production from wind is zero.

Now, to be fair and balanced, the LEGO® organization should next build a nuclear power plant out of LEGOs to show kids the importance of using a non-polluting, reliable source of energy. They could even use biodegradable plastic blocks made from algae just to be accurate. At this time, it is likely that nuclear fusion will be our ultimate energy source. We have time to develop that technology thanks to abundant fossil fuels. However, we will not have the capital to do that if we waste our resources tilting at the Green New Deal windmill now being erected by Democratic Socialists and Marxists (same thing).

An early use of the term Green New Deal was by journalist Thomas L. Friedman. He argued in favor of the idea in two pieces that appeared in the New York Times and The New York Times Magazine. In January 2007, Friedman wrote:

> "If you have put a windmill in your yard or some solar panels on your roof, bless your heart. But we will only green the world when we change the very nature of the electricity grid -- moving it away from dirty coal or oil to clean coal and renewables. And that is a huge industrial project -- much bigger than anyone has told you. Finally, like the New Deal, if we undertake the green version, it has the potential to create a whole new clean power industry to spur our economy into the 21st century."

I am not a fan of Mr. Friedman's positions on many things, but I tend to agree with him on his key points here, in particular, the "clean coal" comment. Unfortunately, there is an implication that this is an urgent matter, *which it isn't*. Urgency leads to a government program as the only solution, which also, *it isn't*. But the seed of urgency has been planted and is being hyped by non-journalistic media. (By non-journalistic I mean unethical and biased, perhaps lazy. It didn't use to be that way.)

This approach was subsequently taken up by the Green New Deal Group, which published its eponymous report on July 21, 2008. The concept was further popularized and put on a wider footing when the United Nations Environment Programme (UNEP) began to promote it. On October 22, 2008 UNEP's Executive Director Achim Steiner unveiled the Global Green New Deal initiative that aims to create jobs in "green" industries, thus boosting the world economy and curbing climate change at the same time.

On January 8, 2019, Mr. Friedman reopened the subject in an New York times article entitled *The Green New Deal Rises Again*.

"Back in 2007, I wrote a column calling for a Green New Deal, and I later expanded on the idea in a book, Hot, Flat and Crowded. Barack Obama picked up the theme and made a Green New Deal part of his 2008 platform, but the idea just never took off. So, I'm excited that the new Democratic Congresswoman Alexandria Ocasio-Cortez and others have put forward their own takes on a Green New Deal, and it's now getting some real attention . . .

Friedman continues his column: "The Green New Deal that Ocasio-Cortez has laid out aspires to power the U.S. economy with 100 percent renewable energy within 12 years and calls for "a job guarantee program to assure a living wage job to every person who wants one," "basic income programs" and "universal health care," financed, at least in part, by higher taxes on the wealthy. Critics argue that this is technically unfeasible and that combining it with democratic socialist proposals will drive off conservatives needed to pass it . . .

My own definition of a Green New Deal, which has evolved since 2007 as the technology has gotten better and the climate problem has gotten worse, remains focused on how a green revolution in America can drive innovation, spur new industries and enhance our security . . .

Who believes that America can remain a great country and not lead the next great global industry? Not me. A Green New Deal, in other words, is a strategy for American national security, national resilience, natural security and economic leadership in the 21st century. Surely some conservatives can support that.

And to make sure that they have an incentive to, I would also guarantee that a portion of every dollar raised by a carbon tax in a Green New Deal would be invested in two new community colleges and high-speed broadband in rural areas of every state. Each state could decide where. Every American needs to feel a chance to gain from a Green New Deal . . .

I am eager to see what other people propose, but we don't have another decade to waste. This may well be our last chance

> to build the technologies we need at the scale of the challenge we face in the time we still have to — as scientists say — manage the unavoidable aspects of climate change and avoid the unmanageable ones . . .
>
> As the environmentalist Dana Meadows once put it, "We have exactly enough time — starting now."

That is really more pages in my book than I wanted to give to these New York Times opinion writers but I also wanted to be fair. These liberal, progressive thinkers, such as Friedman, have done great damage to the Nation and its Founding principles. The damage will be even worse if a Carbon Tax results from their energy ignorance. But the Green New Deal contains the seed of a Carbon Tax now planted in the shallow soils of veritable Socialistic gardens--the West Coast and East Coast corridor. But, as the Great Communicator, President Ronald Reagan once said, "Here you go again." Friedman has laid out the East and West Coast reasoning. The Democratic Socialist lemmings now follow the scent of the Big Lie.

There are only a few people attempting to speak Truth to Power on this subject. One of my favorites is Alex Epstein. I have yet to have a disagreement with any of Alex's positions on the increasing value of fossil fuels with respect to the concept of human flourishing. However, there are many forces arrayed against the notion that the natural resources of the world are for the benefit of humankind. Many Americans are in the process of being convinced by subversive interests that humans are a blight on the world. My suspicion is that the subversive attack is on the institution of Individual Freedom, not fossil fuels. I have written two books on that subject. In this new book, my objective has been to present a reasonable approach to the ultimate objective of clean energy. To get there, some idealistic notions have to be debunked. Let's begin.

# THE BIG LIE

Andy May, author of *CLIMATE CATASTROPHE!: Science or Science Fiction?* explains:

"Alexandria Ocasio-Cortez is only the latest in a long line of politicians to use climate change as an excuse for world government and global control of production, distribution and exchange of goods and services, aka socialism. The global warming (or climate change, if you prefer) scare has been inexorably tied to socialism since it was conceived in the late 1980s by Maurice Strong (see the details of what Strong did in Christopher Booker's article on him in the 5 December 2015 issue of *The Telegraph*). In short, he became the founding director of the UN Environmental Programme (UNEP) and later, in 1992, he created the UN Framework Convention on Climate Change.

The UN Framework Convention on Climate Change has dominated (some would say "dictated") the global climate change agenda ever since. In 2015, the then Executive Secretary of the body, Christiana Figueres, said this:

"This is the first time in the history of mankind that we are setting ourselves the task of intentionally, within a defined period of time to change the economic development model that has been reigning for at least 150 years, since the Industrial Revolution."

Christine Stewart, the Canadian Minister of the Environment said to the editors and reporters of the Calgary Herald in 1998:

"No matter if the science of global warming is all phony ... climate change [provides] the greatest opportunity to bring about justice and equality in the world."

Former Senator Timothy Wirth, when representing the Clinton-Gore administration as Undersecretary of State for Global Affairs in 1993, advocated using the global warming issue to promote global economic change: "We have got to ride the global warming issue. Even if the theory of global warming is wrong, we will be doing the right thing in terms of economic policy and environmental policy."

Thus, the global warming/climate change is not a scientific issue, it is an economic and political one. By speculating that climate change is man-made, through our carbon dioxide emissions, and dangerous, the politicians can claim that to save the planet we must form a global governmental body to reduce carbon dioxide emissions and "save the planet."

However, whether climate change is mostly natural or mostly man-made, is less important than the rate of the change and whether it is dangerous. The rate of global warming over the past 150 years is less than one degree Celsius (1.8°F) per 100 years, this is not alarming and, if anything, it appears to be slowing down in recent decades (Fyfe, et al., 2016). Further, our oceans, which cover 70% of the Earth's surface, provide an upper limit on the Earth's surface temperature of around 30 degrees C according to studies by MIT atmospheric physicists Newell and Dopplick (Newell & Dopplick, 1979). So, a climate catastrophe is not headed our way anytime in the next few hundred years, we do have plenty of time to study the matter. According to a study of the Earth's climate history by Christopher Scotese and the Paleomap Project, at the University of Texas (Scotese, 2015), the average surface of the Earth over the past 500 million years is about 20 degrees C (68°F). This is over five degrees warmer (9°F) than today, thus we are in an unusually cold period in the Earth's history. In the past, the Earth's average surface temperature has been as warm as 28 degrees C (82°F), during these times dinosaurs roamed over the continent of Antarctica and palm trees grew on the North Slope of Alaska, while equatorial temperatures remained about the same as today.

Characterizing the current warming as an urgent and impending crisis is silly considering the scientific evidence we have today. There is no need to remove national boundaries, form a global government and abandon capitalism to "save the world." Climate changes, we all accept this, perhaps it is mostly man-made, perhaps it is mostly natural, we don't know. What we do know is that many communities may be affected by climate change. Sea level is rising, the best long-term estimates are that it is rising between 1.8 and 3 millimeters per year. This is not a large rate, perhaps seven inches to a foot in 100 years, much less than the daily tides. But, if it causes problems, seawalls can be built, people can move from dangerous areas or elevate their houses, it is a problem that can be dealt with locally, as it has been for thousands of years.

With fossil fuels or nuclear power, which the climate change alarmists want to eliminate, we can cool or heat our buildings if a community gets too cold or too hot. If we get more rain, we can improve our drainage or move out of flood plains. If it gets too dry, we can drill wells for water or move water via aqueducts. The point is, each community needs to deal with its own problems. Climate change is not a problem that must be dealt with globally, the people affected and closest to the problem will deal with it in the most effective and efficient manner, as they always have." (May, 2018)

The role of the U. S. Senate was once supposed to be the home of the adults. The Senate was the cooling saucer for the hot liquid spilling from the House of Representatives. *But no more.*

According to Senator Ron Wyden (D) of Oregon, "There is a groundswell of support in America to tackle the calamity of climate change. Addressing climate change is a national security issue, but renewable energy is also a jobs issue, a health issue and a pocketbook issue for each American family. Constantly improving technology means that low-carbon renewable energy is cleaner, cheaper and safer than burning fossil fuels.

Congress needs to kick the carbon habit. We need a 'Green New Deal' that helps all Americans take advantage of this fundamental change — as workers and manufacturers, consumers, builders and inventors. And much like the original New Deal — when President Franklin D. Roosevelt worked with Congress to pass 15 major bills in 100 days — a comprehensive 'Green New Deal' will require swift action on several legislative proposals to meet its goal. As the 116th Congress convenes, one essential legislative proposal it should move on quickly is to throw in the trash the 44 separate energy tax breaks, anchored by advantages for big oil companies that get billions of dollars in beneficial tax treatment.

The dirty relics of the past century should be replaced with just three new energy tax incentives: one for clean energy, one for clean transportation fuel and one for energy efficiency. Under this new system, benefits would be received only if carbon emissions are decreased or eliminated. The cleaner it is, the greater the benefit. These reforms will not only set off

> a wave of investment and innovation in clean and renewable energy, they will also cut subsidies and save Americans money … The U.S. tax code is now stacked against renewable energy; it's stacked against technologies that can lower energy carbon outputs and it's stacked against taxpayers. The Clean Energy for America Act ends these skewed priorities at a time when poll after poll show that a majority of Americans want climate action.
>
> Scientists are sounding the alarms. Citizens are demanding more than the status quo. Urgency is everything. Failure to act spells dire consequences for our economy, the health and safety of our families and most vulnerable communities, and the future of our planet. Let's take a first step for a 'Green New Deal' by bringing the spirit of FDR to fix our broken tax code." (Wyden, 2019)

*Hogwash Senator Wyden.* The tax code is broken but the Green New Deal isn't going to fix it. It will just make things worse. The majority of Americans do not want climate action. Climate Action is a jumbled-up term used by pollsters with hidden agendas in support of the Green Movement. The goal of the Green Movement is the removal of humankind from the face of earth. It is not about Clean Energy or improving the lot of the poor and downtrodden. The young Democratic Socialists are being duped by their own sense of morality.

The biggest problem here is not the idea or the non-factual comments about climate change, carbon, Big Oil, or the concept of "Green Energy" by Senator Wyden. It is the urgency implication in addition to the idea that this must be a State-run program. In government, urgency means mistakes with its associated cost over-runs and poor final results. A case-in-point would be the Affordable Care

Act. That nationalized 'socialized medicine' plan was rushed for political expediency without any journalistic challenges. The Green New Deal is another Bad Deal for America.

Transitioning away from carbon-based fuels must follow a common-sense approach with cost to benefit analysis driving every decision, the very opposite of political behavior. It will not be easy and cannot be hamstrung with partisan political bickering or diluted by vested interests. The proposed Green New Deal contains all these contaminants, but probably the most dangerous to the Nation are the use of Marxist principles such as totalitarian State control of production and controlling the behavior of American citizens by implementing a Carbon Tax.

In a letter published by The Wall Street Journal on January 19, 2019, 45 economists said a Carbon Tax "offers the most cost-effective lever to reduce carbon emissions at the scale and speed that is necessary" to fight climate change.

The new tax would affect the price of goods or services that emit carbon, like fossil fuels. Economists said "energy bills and gasoline prices would rise *at first* [and continue to rise], but they called on the government to return that tax revenue directly to American citizens [but allocated on needs determined by the State]."

This is sheer nonsense and contrary to real economic principles on so many levels. First of all, this will be a tax on the middle-class and there will be no dividend returned, *there never is.* Second , it will not have a significant impact on the climate. Third, it will lower the United States competitive standing in the world economy. And finally, there is no urgency in transitioning away from carbon-based fuels. Any politician of any stripe who buys into the idea of a Carbon Tax has not performed "due diligence" and should not be re-elected.

# PATHOLOGICAL SCIENCE

Global warming, climate change, and the sky is falling. The consensus of the majority of scientists agree, "The sky is falling. The planet is doomed if we humans don't stop breathing and surviving." But hold on, it's not really a consensus of scientists in agreement on some subject. It's a list of research paper titles with conclusions that have been misrepresented. The papers have not been peer reviewed and have been tainted by bias towards the agenda of the source of funding, the United Nations, Progressives in our own government and the fund receivers, left leaning liberal Academia. It is Pathological Science.

Pathological Science is an area of research where "people are tricked into false results ... by subjective effects, wishful thinking or threshold interactions." The term was first used by Irving Langmuir, Nobel Prize–winning chemist, during a 1953 colloquium at the Knolls Research Laboratory. Langmuir said a pathological science is an area of research that simply will not "go away"—long after it was given up on as "false" by the majority of scientists in the field. He called pathological science "the science of things that aren't so." (Pathological Science, 2019)

Pathological science, as defined by Langmuir, is a psychological process in which a scientist, originally conforming to the scientific method, unconsciously veers from that method, and begins a pathological process of wishful data interpretation. Some characteristics of pathological science are:

--The maximum effect that is observed is produced by a causative agent of barely detectable intensity, and the magnitude of the effect is substantially independent of the intensity of the cause. *For example: human-caused carbon dioxide emissions.*

--The effect is of a magnitude that remains close to the limit of detectability, or many measurements are necessary because of the very low statistical significance of the results. *For example: Surface and atmospheric temperature measures and sea level rise observations.*

--There are claims of great accuracy. *For example: 97 percent of all scientists agree. Computer-generated Climate models are always correct.*

--Fantastic theories contrary to experience are suggested. *For example: The Arctic and Antarctic ice is melting rapidly, the seas are rising, glaciers are receding, more hurricanes and severe weather are occurring, catastrophic climate change is upon us. The rise in global average temperature follows increases in atmospheric carbon dioxide produced by burning fossil fuels.*

--Criticisms are met by ad hoc excuses. *For example: The recent leveling off in the rise of average global warming is an anomaly. The fact that the 'hockey stick' graph shows that carbon dioxide increases occur after global temperature increases is subject to interpretation.*

--The ratio of supporters to critics rises and then falls gradually to oblivion. *For example: Everyone agrees that catastrophic climate change is real and is caused by the use of fossil fuels except that catastrophic climate change does not seem to be occurring. Humm.*

So, it seems that the 97 percent consensus of all Pathological Scientists is that humans are causing catastrophic climate change by burning fossil fuels. Imagine that. *We bad.*

# ENERGY TRUTH

According to the EIA, "Energy is the ability to do work and comes in different forms such as Heat (thermal), Light (radiant), Motion (kinetic), Electrical, Chemical, Nuclear energy and Gravitational. People use energy for everything from walking to sending astronauts into space. There are two types of energy: Stored (potential) energy and Working (kinetic) energy. For example, the food a person eats contains chemical energy, and a person's body stores this energy until he or she uses it as kinetic energy during work or play.

When people use electricity in their homes, the electrical power is probably generated by burning coal or natural gas, by a nuclear reaction, or by a hydroelectric plant on a river, to name just a few sources. When people fill up a car's gasoline tank, the energy source is petroleum (gasoline) refined from crude oil and may include fuel ethanol made by growing and processing corn. Coal, natural gas, nuclear, hydropower, petroleum, and ethanol are called energy sources.

Energy sources are divided into two groups: Renewable (an energy source that can be easily replenished) and Nonrenewable (an energy source that cannot be easily replenished). Renewable and nonrenewable energy sources can be used as primary energy sources to produce useful energy such as heat or used to produce secondary energy sources such as electricity [and hydrogen].

According to the EIA, there are five main renewable energy sources: Solar energy from the sun, Geothermal energy from heat inside the earth, Wind energy, Biomass from plants and Hydropower from flowing water.

The nonrenewable energy sources are petroleum products, hydrocarbon gas liquids, natural gas, coal and nuclear energy. Today, most of the energy consumed in the United States is from nonrenewable energy sources. Crude oil, natural gas, and coal are called fossil fuels because they were formed over millions of years by the action of heat from the earth's core and pressure from rock and soil on the remains (or fossils) of dead plants and creatures such as microscopic diatoms. Most of the petroleum products consumed in the United States are made from crude oil, but petroleum liquids can also be made from natural gas and coal.

Nuclear energy is produced from uranium, a nonrenewable energy source whose atoms are split (through a process called nuclear fission) to create heat and, eventually, electricity." (EIA, 2019)

The chart below shows the energy sources used in the United States. In 2017, nonrenewable energy sources accounted for about 90% of U.S. energy consumption in 2017. Biomass, which includes wood, biofuels, and biomass waste, is the largest renewable energy source, and it accounted for nearly half of all renewable energy consumption and about 5% of total U.S. energy consumption.

## U.S. energy consumption by energy source, 2017

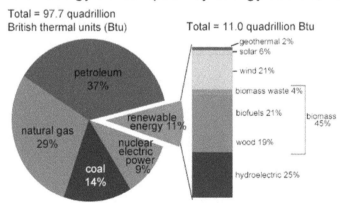

Total = 97.7 quadrillion British thermal units (Btu)

Total = 11.0 quadrillion Btu

Note: Sum of components may not equal 100% because of independent rounding.
Source: U.S. Energy Information Administration, *Monthly Energy Review*, Table 1.3 and 10.1, April 2018, preliminary data

# THE BIG GREEN PROBLEM

The Green New Deal sounds great. Initiate a National program along the lines of The Manhattan Project to reduce, even eliminate greenhouse gas emissions in the United States, then show the world how to do just that. The purpose for reducing greenhouse gas emissions is to prevent runaway global warming that will cause catastrophic climate change.

But wait, here's the problem with all of this. Increases in greenhouse gases that are attributable to human activity are miniscule and have a logarithmic limited impact on average global temperatures. Average global temperature has little to do with climate change. It's the differences in temperature at various locations around the earth and the earth atmosphere that cause weather changes. If the new temperature at each location remains stagnant for a while, the climate will then change or adjust. This has been going on for eons and has very little to do with human-caused carbon dioxide emissions. Weather variations are caused by uneven heating of the earth surface (which includes the oceans) and the atmosphere by the sun. The nature of this uneven heating is cyclical. Short changes in heating cause the weather, longer changes cause weather conditions that control the climate at each unique location on the earth's surface.

From a non-technical standpoint, it seems logical to ponder the causes of the uneven heating of the earth surface and atmosphere. Solar output varies, somewhat on a cyclical basis. There are anomalies in this cycle which we have observed as solar storms that are visible as sun spots but there seems to be a measurable frequency to these solar storms. The earth orbits the sun on a regular basis, but it varies in its distance to the sun because the orbit is an ellipse, not a circle. The earth also wobbles in its rotation which causes uneven heating. These mechanisms which result in the uneven heating of the earth are the primary causes of climate change. The climate changes, it always will.

With the understanding that it is uneven heating that causes weather and climate change, then what is the value of reducing the amount of human-controllable greenhouse gas, primarily carbon dioxide from burning fossil fuel? Actually none, zero, nada. Carbon dioxide is not toxic so there is no environmental impact other than to improve the growth of green plants, and that's a good thing. But controlling the toxic emissions from the combustion of fossil fuels is also a good thing. That is already being done.

Reducing greenhouse gas emissions, the declared objective of the Green New Deal, is the wrong objective. The objective should be to develop a plan for the next era of Energy, a plan that will ease the transition away from carbon-based fuels and towards a cleaner, more abundant energy source that will be economically viable. The plan should utilize the available energy sources as a bridge to the ultimate source, which may very well turn out to be human controlled, portable nuclear fusion. The desire for the energy source to be portable is to allow its use for transportation, perhaps interstellar, who knows? However, to realistically get there, we must make optimum use of available energy sources, whether they are renewable or not. The most affordable source today is carbon-based fuel in the form of petroleum, natural gas and coal. It is affordable because it is readily available in vast amounts, it is not intermittent or dilute, and the current human infrastructure was designed for its use. This energy availability will last for more than a century or two so there is time to achieve the ultimate goal without destroying the reasons for the very existence of the United States of America--Life, Liberty and the Pursuit of Happiness defined as Individual Freedom. Let's begin.

# PEAK RENEWABLE ENERGY

"Renewable energy is energy from sources that are naturally replenishing but flow-limited. They are virtually inexhaustible in duration but limited in the amount of energy that is available per unit of time. The major types of renewable energy sources are Biomass which includes wood and wood waste, municipal solid waste, landfill gas and biogas, ethanol, biodiesel, hydropower, geothermal, wind and solar. Until the mid-1800s, wood was the source of nearly all of the nation's energy needs for heating, cooking, and light. From the late 1800's until today, fossil fuels—coal, petroleum, and natural gas—have been the major sources of energy. Hydropower and solid biomass were the most used renewable energy resources until the 1990s. Since then, the shares of U.S. energy consumption from biofuels, solar, and wind energy have increased.

In 2017, renewable energy provided about 11 quadrillion British thermal units (Btu) 11 quadrillion British thermal units (Btu)—1 quadrillion is the number 1 followed by 15 zeros—equal to 11% of total U.S. energy consumption. About 57% of U.S. renewable energy consumption was by the electric power sector, and about 17% of U.S. electricity generation was from renewable energy sources." (EIA, 2019)

According to the EIA, "Renewable energy plays an important role in reducing greenhouse gas emissions. Using renewable energy can reduce the use of fossil fuels, which are major sources of U.S. carbon dioxide emissions." But does it really? The major reduction in carbon dioxide emissions for the last few years has been due to the conversion of coal-fired power plants to natural gas-fired plants. This has also led to great improvements in the reduction of toxic emissions. However, technology to reduce emissions from the combustion of coal is also advancing and will continue to make coal competitive as an energy source in the world, not just the U. S., for many years to come.

The consumption of biofuels and other nonhydroelectric renewable energy sources more than doubled from 2000 to 2017, mainly because of state and federal government requirements and incentives to use renewable energy. The U.S. Energy Information Administration projects that "U.S. renewable energy consumption will continue to increase through 2050." (EIA, 2019)

Now let's get serious. In 2017, renewables only supplied 11 percent of the U. S. energy demand and that includes hydroelectric plants. The Green New Deal, with its Green Movement supporters, will not allow an increase in hydroelectric plants (dams) since they are considered "non-Green." The largest remaining component is Biomass. It is 45% of the renewable mix but it is also "non-Green." Most of the Biomass energy involves combustion with some emissions somewhere along the line. Furthermore, 21% of the Biomass energy is from Biofuels. Biofuels impact the food supply and require major subsidies. The most likely successful biofuel may eventually come from algae, not the plant food supply. For obvious reasons, biofuel from algae will certainly be Green, but not abundant for a while.

That leaves Wind and Solar as the only, Green Movement-approved renewable energy sources. These intermittent, dilute energy sources are supposed to provide 100% of all U. S. energy requirements within 12 years according to the goal of the proposed Green New Deal draft (or daft) legislation. Solar and Wind currently supply less than 3 % of U. S. Energy demand and both rely on backup nonrenewable power supplies and subsidies to be operationally and economically viable. To think that Solar and Wind will provide 100% of U. S. energy requirements in 12 years is sheer nonsense.

There may be a way to determine the ultimate level of energy that could be supplied by Wind and Solar in the U. S. This would involve some projections on wind patterns, cloud patterns, areal extent (wind and solar farms), manufacturing ability, material resources and such. But, technically speaking, there is a limit for Wind and Solar.  It will be well below 100%. With my wet engineering finger in the air, I will say maybe 20% max (and that is generous).  Prove me wrong.

## WIND ENERGY

But back to serious thought now. Let's have a look at energy from wind. "Wind is caused by uneven heating of the earth's surface by the sun." Good to know there is uneven heating. "Because the earth's surface is made up of different types of land and water [and ice], it absorbs the sun's heat at different rates. One example of this uneven heating is the daily wind cycle. During the day, air above the land heats up faster than air over water. Warm air over land expands and rises, and heavier, cooler air rushes in to take its place, creating wind. At night, the winds are reversed because air cools more rapidly over land than it does over water. In the same way, the atmospheric winds that circle the earth are created because the land near the earth's equator is hotter than the land near the North Pole and the South Pole." (EIA, 2019)

Wind energy is not new. Windmills have been used to grind grain, pump water for irrigation and flood control, and some manufacturing (sawmills for example). Water-pumping windmills were once used throughout the United States and some still operate on farms and ranches, mainly to supply water for livestock. Today, wind energy is mainly used to generate electricity.

Wind turbines use blades to collect the wind's kinetic energy. Wind flows over the blades creating lift (similar to the effect on airplane wings), which causes the blades to turn. The blades are connected to a drive shaft that turns an electric generator, which produces the electricity.

In 2017, wind turbines in the United States were the source of nearly 6.3% of total U.S. utility-scale electricity generation. Electricity generation from wind in the United States increased from about 6 billion kilowatt-hours (kWh) in 2000 to about 254 billion kWh in 2017.

New technologies have decreased the cost of producing electricity from wind but most of the growth in wind power has been due to government subsidies and other artificial incentives.

Operating a wind power plant is more complex than simply erecting wind turbines in a windy area. Wind power plant owners must carefully plan where to position wind turbines and must consider how fast and how often the wind blows at the site. An individual wind turbine has a relatively small physical footprint. Wind farms are located on open land, on mountain ridges, or offshore in lakes or the ocean.

Wind turbines do have some negative effects on the environment. Modern wind turbines can be very large machines, and they may visually affect the landscape. A small number of wind turbines have also caught fire, and on rare occasions have leaked lubricating fluids. Some wind turbine blades are noisy while other types of wind turbines and wind projects cause bird and bat deaths.

Most wind power projects on land require service roads that add to the physical effects on the environment. Wind turbines may also use rare earth minerals. These minerals are often located in countries with less stringent environmental standards than the U. S. and mining these minerals can have negative effects on the environment. Producing the metals and other materials used to make wind turbines and the concrete used for their foundations requires energy that has most likely been provided by carbon-based fuels.

## SOLAR ENERGY

The amount of solar energy that the earth receives each day is many times greater than the total amount of all energy that people consume each day. However, on the surface of the earth, solar energy is a variable and intermittent energy source. The amount of sunlight and the intensity of sunlight varies by time of day and location. Weather and climate conditions affect the availability of sunlight daily and on a seasonal basis. The type and size of a solar energy collection and conversion system determines how much of the available solar energy can be converted into useful energy.

## SOLAR THERMAL COLLECTORS

There are many practical ways to capture solar energy. The most basic are solar thermal collectors. Low-temperature solar thermal collectors absorb the sun's heat energy to heat water for washing and bathing or for swimming pools, or to heat air inside buildings.

Concentrating collectors are more complex. Concentrating solar energy technologies use mirrors to reflect and concentrate sunlight onto receivers that absorb solar energy and convert it to heat. We use this thermal energy for heating homes and buildings or to produce electricity with a steam turbine or a heat engine that drives a generator. All solar thermal power systems have solar energy collectors with two main components. The first component consists of reflectors (mirrors) that capture and focus sunlight onto a receiver and a heat-transfer fluid. The second component consists of a boiler and steam driven turbine that powers a generator to produce electricity. Solar thermal power systems have tracking systems that keep sunlight focused onto the receiver throughout the day as the sun changes position in the sky.

Solar thermal power systems may also have a thermal energy storage system component that allows the solar collector system to heat an energy storage system during the day, and the heat from the storage system is used to produce electricity in the evening or during cloudy weather. Solar thermal power plants may also be hybrid systems that use other fuels (usually natural gas) to supplement energy from the sun during periods of low solar radiation. These types of systems are usually very large and operated by utility companies or government installations.

For private and limited commercial-use, the two general types of solar heating systems are passive systems and active systems. Passive solar space heating happens when the sun shines through the windows of a building and warms the interior. Building designs that optimize passive solar heating usually have south-facing windows that allow the sun to shine on solar heat-absorbing walls or floors during the winter. The solar energy heats the building by natural radiation and convection. Window overhangs or shades block the sun from entering the windows during the summer to keep the building cool.

Active solar heating systems use a collector and a fluid that absorbs solar radiation. Fans or pumps circulate air or heat-absorbing liquids through collectors and then transfer the heated fluid directly to a room or to a heat storage system. Active solar water heating systems usually have a tank for storing solar heated water.

Solar collectors are either non-concentrating or concentrating. In non-concentrating collectors the collector area (the area that intercepts the solar radiation) is the same as the absorber area (the area absorbing the radiation). Solar systems for heating water or air usually have non-concentrating collectors. Flat-plate collectors are the most common type of non-concentrating collectors for water and space heating in buildings and are used when temperatures lower than 200°F are sufficient.

Solar water heating collectors have metal tubes attached to the absorber. A heat-transfer fluid is pumped through the absorber tubes to remove heat from the absorber and transfer the heat to water in a storage tank. Solar systems for heating swimming pool water in warm climates usually do not have covers or insulation for the absorber, and pool water is circulated from the pool through the collectors and back to the pool. Solar air heating systems use fans to move air through flat-plate collectors and into the interior of buildings.

## SOLAR PANELS

Solar energy can also be captured with photovoltaic systems. This is the method most people think of when the discussion is about solar energy. Photovoltaic (PV) cells convert sunlight directly into electricity. PV systems can range from systems that provide tiny amounts of electricity for watches and calculators to systems that provide the amount of electricity that hundreds of homes now use.

Millions of houses and buildings around the world now have PV systems on their roofs. Many multi-megawatt PV power plants have also been built. Covering 4% of the world's desert areas with photovoltaics could supply the equivalent of all of the world's daily electricity use in 2017. (EIA, 2019)

Solar energy systems/power plants do not produce air pollution, water pollution, or greenhouse gases. Using solar energy can have a positive, indirect effect on the environment when solar energy replaces or reduces the use of other energy sources that have larger effects on the environment.

However, some toxic materials and chemicals are used to make the photovoltaic (PV) cells that convert sunlight into electricity. Some solar thermal systems use potentially hazardous fluids to transfer heat. Leaks of these materials could be harmful to the environment. U.S. environmental laws regulate the use and disposal of these types of materials.

As with any type of power plant, large solar power plants can affect the environment near their locations. Clearing land for construction and the placement of the power plant may have long-term effects on the habitats of native plants and animals. Some solar power plants may require water for cleaning solar collectors and concentrators or for cooling turbine generators. Using large volumes of ground water or surface water in some arid locations may affect the ecosystems that depend on these water resources. In addition, the beam of concentrated sunlight a solar power tower creates can kill birds and insects that fly into the beam.

# THE REAL PROBLEM WITH WIND AND SOLAR

Wind and solar energy sources are unreliable because they are by definition intermittent. They are also dilute and require substantial resources, many of the non-renewable type, just to be created and to operate. The resulting product, electrical energy, is more costly per unit that that provided by alternative sources. And, for the U. S., there is a finite limit on its ultimate contribution to meet demand.

In 2016, the EIA announced that: "Wind, natural gas, and solar made up almost all new electric generation capacity in 2015, accounting for 41%, 30%, and 26% of total additions, respectively, according to preliminary data. The data also show a record amount of distributed solar photovoltaic (PV) capacity was added on rooftops throughout the country in 2015." Keep in mind, that it is always going to be a record amount even if the increase is trivial.

Such statistics are cited in the US and around the world by anti-fossil/anti-nuclear/anti-hydro Green groups to argue that their policies won't lead to energy poverty but rather a future full of cheap, plentiful, reliable 'solar and wind energy'.

Alex Epstein's take on this is that "…extravagant claims always use a misleading word that, if you spot, you know something deceptive is going on. That word is capacity–as in wind being the leading new source of 'electrical generation capacity.' When you hear that wind has the most increased capacity, you are supposed to think that it has the most increased ability to provide electricity in the way we need it–affordably and reliably.

But in energy, 'capacity' is actually a technical term meaning the maximum momentary ability to produce electricity–not the consistent, long-term ability to produce electricity, which is what matters to human life.

For the kinds of energy I call 'reliables'–coal, oil, gas, hydro, nuclear, capacity is roughly equal to ability because their fuel sources are stored, always available, and therefore controllable. A nuclear power plant, for example, might have the ability to run at 90% of 'capacity' month after month.

But for the kinds of energy I call 'unreliables'–solar and wind, whose fuel sources are intermittent, unpredictable, and most of the time unavailable, the term 'capacity' is inherently misleading. A wind farm may operate near maximum capacity at brief, unpredictable moments and produce little to nothing the rest of the time. Those unreliable bursts might add up to 20-30% its supposed capacity. A set of solar panels may operate near capacity in the middle of the summer in the middle of the day when there are no clouds, but most of the time it has far less ability, when clouds (or non-summer seasons) come that ability can disappear, and at night the panels obviously have no electrical generation ability. For the purposes of providing individuals the cheap, plentiful, on-demand electricity they need, this is useless.

The actual ability of wind and solar is essentially zero. Witness the celebrated electric grid of Germany. Since solar and wind can always dip to zero, citizens of Germany must purchase enough real capacity from reliables to give everyone the electricity they need. Thus, the solar and wind are unnecessary and indeed problematic since they add unpredictable, destabilizing electricity to the grid. Such wastefulness helps explain why Germans pay 3-4 times for electricity what we do in the US." [Some can't afford it and a new term "Energy Poverty" is now in use in Germany]. "Every time you hear some claim about wind and solar capacity remember that since their reliable capacity is zero, more 'capacity' means more dead weight and higher prices–until and unless someone can create independent solar or wind power plants with an affordable mass storage system. So far, there isn't one." (Epstein, Why Green Energy Means No Energy, 2019)

## THE GERMAN EXPERIMENT

"More people are finally beginning to realize that supplying the world with sufficient, stable energy solely from sun and wind power will be impossible. Germany took on that challenge, to show the world how to build a society based entirely on "green, renewable" energy. It has now hit a brick wall. Despite huge investments in wind, solar and biofuel energy production capacity, Germany has not reduced $CO_2$ emissions over the last ten years. However, during the same period, its electricity prices have risen dramatically, significantly impacting factories, employment and poor families."

The German experiment needs to be reviewed in depth by U. S. Energy Planners. "Germany took on the challenge, to show the world how to build a society based entirely on "green, renewable" energy. It has now hit a brick wall. Despite huge investments in wind, solar and biofuel energy production capacity, Germany has not reduced $CO_2$ emissions over the last ten years. However, during the same period, its electricity prices have risen dramatically, significantly impacting factories, employment and poor families.

Germany has installed solar and wind power to such an extent that it should theoretically be able to satisfy the power requirement on any day that provides sufficient sunshine and wind. However, since sun and wind are often lacking – in Germany even more so than in other countries like Italy or Greece – the country only manages to produce around 27% of its annual electric power needs from these sources.

Equally problematical, when solar and wind production are at their maximum, the wind turbines and solar panels often overproduce – that is, they generate more electricity than Germany needs at that time – creating major problems in equalizing production and consumption. If the electric power system's frequency is to be kept close to 50Hz (50 cycles per second), it is no longer possible to increase the amount of solar and wind production in Germany without additional, costly measures.

Production is often too high to keep the network frequency stable without disconnecting some solar and wind facilities. This leads to major energy losses and forced power exports to neighboring countries ('load shedding') at negative electricity prices, below the cost of generating the power.

In 2017 about half of Germany's wind-based electricity production was exported. Neighboring countries typically do not want this often, unexpected power and the German power companies must therefore pay them to get rid of the excess. German customers have to pick up the bill.

If solar and wind power plants are disconnected from actual need in this manner, wind and solar facility owners are paid as if they had produced 90% of rated output. The bill is also sent to customers.

When wind and solar generation declines, and there is insufficient electricity for everyone who needs it, Germany's utility companies also have to disconnect large power consumers – who then want to be compensated for having to shut down operations. That bill also goes to customers all over the nation.

Power production from the sun and wind is often quite low and sometimes totally absent. This might take place over periods from one day to ten days, especially during the winter months.

Conventional power plants (coal, natural gas and nuclear) must then step in and deliver according to customer needs. Hydroelectric and biofuel power can also help, but they are only able to deliver about 10% of the often very high demand, especially if it is really cold. Alternatively, Germany may import nuclear power from France, oil-fired power from Austria or coal power from Poland.

In practice, this means Germany can never shut down the conventional power plants, as planned. These power plants must be ready and able to meet the total power requirements at any time; without them, a stable network frequency is unobtainable. The same is true for French, Austrian and Polish power plants.

Furthermore, if the AC frequency is allowed to drift too high or too low, the risk of extensive blackouts becomes significant. That was clearly demonstrated by South Australia, which also relies heavily on solar and wind power, and suffered extensive blackouts that shut down factories and cost the state billions of dollars.

The dream of supplying Germany with mainly green energy from sunshine and wind turns out to be nothing but a fading illusion. Solar and wind power today covers only 27% of electricity consumption and only 5% of Germany's total energy needs, while impairing reliability and raising electricity prices to among the highest in the world.

However, the Germans are not yet planning to end this quest for utopian energy. They want to change the entire energy system and include electricity, heat and transportation sectors in their plans. This will require a dramatic increase in electrical energy and much more renewable energy, primarily wind.

To fulfill the German target of getting 60% of their total energy consumption from renewables by 2050, they must multiply the current power production from solar and wind by a factor of

15. They must also expand their output from conventional power plants by an equal amount, to balance and backup the intermittent renewable energy. Germany might import some of this balancing power, but even then, the scale of this endeavor is enormous.

Perhaps more important, the amount of land, concrete, steel, copper, rare earth metals, lithium, cadmium, hydrocarbon-based composites and other raw materials required to do this is astronomical. None of those materials is renewable, and none can be extracted, processed and manufactured into wind, solar or fossil power plants without fossil fuels. This is simply not sustainable or ecological.

Construction of solar and wind 'farms' has already caused massive devastation to Germany's wildlife habitats, farmlands, ancient forests and historic villages. Even today, the northern part of Germany looks like a single enormous wind farm. Multiplying today's wind power capacity by a factor 10 or 15 means a 200-meter high (650-foot-tall) turbine must be installed every 1.5 km (every mile) across the entire country, within cities, on land, on mountains and in water. In reality, it is virtually impossible to increase production by a factor of 15, as promised by the plans.

The cost of Germany's 'Energiewende' (energy transition) is enormous: some 200 billion euros by 2015 – and yet with minimal reduction in $CO_2$ emission. In fact, coal consumption and $CO_2$ emissions have been stable or risen slightly the last seven to ten years. In the absence of a miracle, Germany will not be able to fulfill its self-imposed climate commitments, not by 2020, nor by 2030.

What applies to Germany also applies to other countries that now produce their electricity primarily with fossil or nuclear power plants. To reach development comparable to Germany's, such countries will be able to replace only about one quarter of their fossil and nuclear power, because these power plants must remain in operation to ensure frequency regulation, balance and back-up power.

Back-up power plants will have to run idle (on 'spinning reserve') during periods of high output of renewable energy, while still consuming fuel almost like during normal operation. They always have to be able to step up to full power, because over the next few hours or days solar or wind power might fail. So, they power up and down many times per day and week.

The prospects for reductions in $CO_2$ emissions are thus nearly non-existent. Indeed, the backup coal or gas plants must operate so inefficiently in this up-and-down mode that they often consume more fuel and emit more (plant-fertilizing) carbon dioxide than if they were simply operating at full power all the time, and there were no wind or solar installations.

There is no indication that world consumption of coal will decline in the next decades. Large countries in Asia and Africa continue to build coal-fired power plants, and more than 1,500 coal-fired power plants are in planning or under construction.

This will provide affordable electricity 24/7/365 to 1.3 billion people who still do not have access to electricity today. Electricity is essential for the improved health, living standards and life spans that these people expect and are entitled to. To tell them fears of climate change are a more pressing matter is a violation of their most basic human rights." (Lundseng, Johnsen, & Bergsmark, 2019)

# THE REAL GREEN MOVEMENT AGENDA

Nuclear and hydropower are not carbon-based and therefore should be the obvious choices to champion by those concerned with reducing $CO_2$ emissions. But the biggest opponent by far of both of these technologies is the Green Movement.

According to Alex Epstein: "That movement keeps insisting, against all evidence, that their anti-fossil, anti-nuclear, anti-hydro stance is not a problem because solar and wind, unreliable, parasitical sources of energy that increase costs wherever they are significantly deployed, will somehow save the day.

Why does the green movement oppose every practical form of energy?

There is only one answer that can explain this. Greens oppose every practical form of energy not out of love for the non-existent virtues of solar and wind energy, but because they believe practical energy is inherently immoral. It's in their philosophical DNA.

To 'be green' means to minimize our impact on nature. In the green philosophy, the standard of value, the metric by which we measure good and bad is: human nonimpact—does an action make our environment more or less altered by humans? If we take that idea seriously, then practical energy is not a good thing.

Energy is 'the capacity to do work' that is, the capacity to alter the placement of matter in nature from where it is to where we want it to be–to impact it. The fundamental use of energy is to power the machines that transform our environment to meet our needs. If an unaltered, untransformed environment is our standard of value, then nothing could be worse than cheap, plentiful, reliable energy. A consistent advocate of green energy therefore would oppose fossil fuels under any circumstances—if they created no waste, including no $CO_2$, if they were even cheaper, if they would last practically forever, if there were no resource-depletion concerns.

Could this really be true? Yes, in fact history proved it true in the late 1980s. For many decades, the ultimate energy fantasy has been what's called nuclear fusion. Conventional nuclear power is called nuclear fission, which unleashes power through the decay of heavy atoms such as uranium. Nuclear fusion unleashes far more power through fusion of two light atoms, of hydrogen, for example. Fusion is what the sun uses for energy. But all human attempts at fusion so far have been inefficient—they take in more energy than they produce. But if it could be made to work, it would be the cheapest, cleanest, most plentiful energy source ever created. It would be like the problem-free fossil fuels I said the Green leaders would oppose.

If an unaltered, untransformed environment is our standard of value, then nothing could be worse than cheap, plentiful, reliable energy. A consistent advocate of green energy therefore would oppose fossil fuels under any circumstances—if they created no waste, including no $CO_2$, if they were even cheaper, if they would last practically forever, if there were no resource-depletion concerns.

56

In the late 1980s, some reports that fusion was close to commercial reality got quite a bit of press. Reporters interviewed some of the world's environmental thought leaders to ask them what they thought of fusion—testing how they felt, not about energy's human-harming risks and wastes but its pure transformative power. What did they say?

There are some quotes from a story in the Los Angeles Times called 'Fear of Fusion: What if It Works?' Leading environmentalist Jeremy Rifkin: 'It's the worst thing that could happen to our planet.'

Paul Ehrlich: Developing fusion for human beings would be 'like giving a machine gun to an idiot child.'

Amory Lovins was already on record as saying, 'Complex technology of any sort is an assault on human dignity. It would be little short of disastrous for us to discover a source of clean, cheap, abundant energy, because of what we might do with it.'

He is talking here about something that, if it had worked, would have been able to empower every single individual on the globe and that undoubtedly would have given him a longer life through the increased scientific and technological progress a fusion-powered society would make. He's talking about something that could take someone who had never had access to a lightbulb for more than an hour, and give him all the light he needed for the rest of his life.

That is what Amory Lovins regards as disastrous 'because of what we might do with it.' Well, we've seen what we do with energy—we make our lives amazing. We go from physically helpless to physical supermen. We build skyscrapers and hospitals. We take vacations and go on honeymoons. We visit our families and tour the world. We relieve drought and vanquish disease. We transform the planet for the better.

Better—by a human standard of value.

But if your standard of value is unaltered nature, then Lovins is right to worry. With more energy, we have the ability to alter nature more, and we will do so—because transforming our environment, transforming nature, is our means of survival and flourishing."" (Epstein, Why Green Energy Means No Energy, 2019)

It seems clear that the Green Movement Agenda is not what it seems to be on the surface. The only way that a pristine planet can be achieved is by removing the contaminant called Humankind. Large masses of this human pollutant have been removed in the past under the same banner that the Democratic Socialists in American now tout, Socialism for one and all. Take from the wealthy and the big corporations and divide it among those less fortunate (or envious, lazy or easily misguided?)

Sir Winston Churchill had a great take on this:

"Socialism is the philosophy of failure, the creed of ignorance, and the gospel of envy." —Perth, Scotland, 28 May 1948, in Churchill, Europe Unite: Speeches 1947 & 1948 (London: Cassell, 1950), 347.

AND

"The inherent vice of capitalism is the unequal sharing of blessings. The inherent virtue of Socialism is the equal sharing of miseries." —House of Commons, 22 October 1945.

# THE ENERGY PLAN

Let's take off where Alex Epstein left off. This does not have to be complicated but does require some non-emotional thinking. First, throw away the concept of renewable and nonrenewable. These terms are meaningless when the timeframes are so significant and the objective is to move from one form of energy source to another, better source of energy. We do need to define what we mean by better. It should improve the lives of humans. It should be cost effective. It should be a clean source of energy which we will define as having a low impact on the natural environment of the earth and may even improve the environment.

We are seeking an energy source that will most likely be a type of fuel that provides electricity through some means. We would like for the energy source to be scalable to meet personal needs (residential and transport), major industrial and economic requirements and military needs. If it meets personal and economic needs it will obviously be socially justifiable, but that's not the objective.

## THE AVAILABLE RESOURCES

We are going to confine our thoughts to the needs of the United States of America. An easy way to do that is to limit the resources considered to those available to the domestic U. S. The largest source of energy is coal. Next is natural gas then crude oil. Then we have nuclear, hydropower and geothermal. Then biomass, wind and solar. We currently have a surplus of coal, natural gas and oil since we are exporting these fuels. Nuclear is more complicated. Hydropower and geothermal have theoretical limits. Biomass currently has an adverse impact on the food supply but algae development may counter that.

In the United States, about 64% of total electricity generation in 2017 was produced from fossil fuels (coal, natural gas, and petroleum), materials that come from plants (biomass), and municipal and industrial wastes. The substances that occur in combustion gases when these fuels are burned include carbon dioxide, carbon monoxide, sulfur dioxide, nitrogen oxides, particulate matter and heavy metals such as mercury. Nearly all combustion byproducts have negative effects on the environment and human health. Many mitigation methods are being used and are becoming more successful each year to the point that air, soil and water pollution in the U. S. is nearing its lowest historical level since the Industrial Revolution got underway.

**THE ENERGY BRIDGE TO A FUTURE OF CLEAN, AFFORDABLE AND RELIABLE ENERGY WILL BE PROVIDED BY CARBON-BASED FUELS, COAL, CRUDE OIL AND NATURAL GAS.**

## COAL

Coal consumption for energy in 2017 was 14% of the total. It was also the second-largest energy source for U.S. electricity generation in 2017—about 30%. Nearly all coal-fired power plants use steam turbines. A few coal-fired power plants convert coal to a gas for use in a gas turbine to generate electricity. In 2017, coal provided the largest electrical generation share in 18 states, down from 28 states in 2007. (EIA, 2019) Part of the steep decline was artificially driven by onerous environmental regulations initiated as part of what has been labeled as the Obama Administration's "War on Coal." However, another factor, related to environmental regulations but primarily economic, was the increase in natural gas availability. This stemmed from the major technological improvements in natural gas production by the private sector energy industry. The improvements were the combination of horizontal drilling and hydraulic fracturing of shale rock formations. Natural gas is considered a cleaner fuel and natural gas-fired turbine generators are more flexible that coal-fired steam-turbine power plants. For the United States as a whole, coal provided 30% of total electricity generation in 2017. (EIA, 2019)

According to the EIA, the amount of coal that exists in the United States is difficult to estimate because it is buried underground. The most comprehensive national assessment of U.S. coal resources was published by the U.S. Geological Survey (USGS) in 1975, which indicated that as of January 1, 1974, coal resources in the United States totaled 4 trillion short tons. Although more recent regional assessments of U.S. coal resources have been conducted by the USGS, a new national-level assessment of U.S. coal resources has not been conducted.

The EIA publishes three measures of how much coal is left in the United States, which are based on various degrees of geologic certainty and on the economic feasibility of mining the coal.

EIA's estimates for the amount of coal reserves as of January 1, 2018, by type of reserve. The Demonstrated Reserve Base (DRB) is the sum of coal in both measured and indicated resource categories of reliability. The DRB represents 100% of the in-place coal that could be mined commercially at a given time. EIA estimates the DRB at about 475 billion short tons, of which about 69% is underground mineable coal.

Estimated recoverable reserves include only the coal that can be mined with today's mining technology after considering accessibility constraints and recovery factors. EIA estimates U.S. recoverable coal reserves at about 253 billion short tons, of which about 58% is underground mineable coal.

Recoverable reserves at producing mines are the amount of recoverable reserves that coal mining companies report to EIA for their U.S. coal mines that produced more than 25,000 short tons of coal in a year. EIA estimates these reserves at about 16 billion short tons of recoverable reserves, of which 68% is surface mineable coal.

Based on U.S. coal production in 2017 of about 0.78 billion short tons, the recoverable coal reserves would last about 325 years, and recoverable reserves at producing mines would last about 26 years. The actual number of years that those reserves will last depends on changes in production and reserves estimates. (U. S. Coal Reserves, 2019)

**CONCLUSION: COAL IN THE FORM OF CLEAN COAL WILL BE A SUBSTANTIAL COMPONENT OF THE ENERGY MIX WELL BEYOND 2050.**

## CRUDE OIL

In 2017, the United States produced an average of about 9.35 million barrels per day (b/d) of crude oil. But how much crude oil do we have and how long will it last? It really depends on the method of measurement and the category of the reserve and/or resource. Oil and gas reserves are defined as volumes that will be commercially recovered in the future. Reserves are separated into three categories: proved, probable, and possible. To be included in any reserve's category, all commercial aspects must have been addressed, which includes government consent. (EIA, 2019)

Proven reserves are those reserves claimed to have a reasonable certainty (normally at least 90% confidence) of being recoverable under existing economic and political conditions, with existing technology. Proven reserves are further subdivided into "proven developed" (PD) and "proven undeveloped" (PUD). PD reserves are reserves that can be produced with existing wells and perforations, or from additional reservoirs where minimal additional investment (operating expense) is required. PUD reserves require additional capital investment (e.g., drilling new wells) to bring the oil to the surface.

The Proven oil reserves in the United States were 36.4 billion barrels ($5.79 \times 10^9$ m3) of crude oil as of the end of 2014, excluding the Strategic Petroleum Reserve. (EIA, 2019)

Unproven reserves are based on geological and/or engineering data similar to that used in estimates of proven reserves, but technical, contractual, or regulatory uncertainties preclude such reserves being classified as proven. Unproven reserves may be used internally by oil companies and government agencies for future planning purposes but are not routinely compiled. They are sub-classified as probable and possible.

63

-Probable reserves are attributed to known accumulations and claim a 50% confidence level of recovery.

-Possible reserves are attributed to known accumulations that have a less likely chance of being recovered than probable reserves. This term is often used for reserves which are claimed to have at least a 10% certainty of being produced. Reasons for classifying reserves as possible include varying interpretations of geology, reserves not producible at commercial rates, uncertainty due to reserve infill (seepage from adjacent areas) and projected reserves based on future recovery methods.

The Energy Information Administration estimates US undiscovered, technically recoverable oil resources to be an additional 198 billion barrels. Services under the U.S. Department of the Interior estimate the total volume of undiscovered, technically recoverable oil in the United States to be roughly 134 billion barrels. (EIA, 2019)

## WHERE ARE THEY?

### Onshore Reserves

The United States Geological Survey (USGS) estimates undiscovered technically recoverable crude oil Onshore in United States to be 48.5 billion barrels. The last comprehensive National Assessment was completed in 1995. Since 2000 the USGS has been re-assessing basins of the U.S. that are considered to be priorities for oil and gas resources. Since 2000, the USGS has re-assessed 22 priority basins, and has plans to re-assess 10 more basins. These 32 basins represent about 97% of the discovered and undiscovered oil and gas resources of the United States. The three areas considered to hold the most amount of oil are the coastal plain (1002) area of ANWR, the National Petroleum Reserve of Alaska, and the Bakken Formation.

Figure 2. Proved reserves of the top seven U.S. oil reserves states, 2013–17

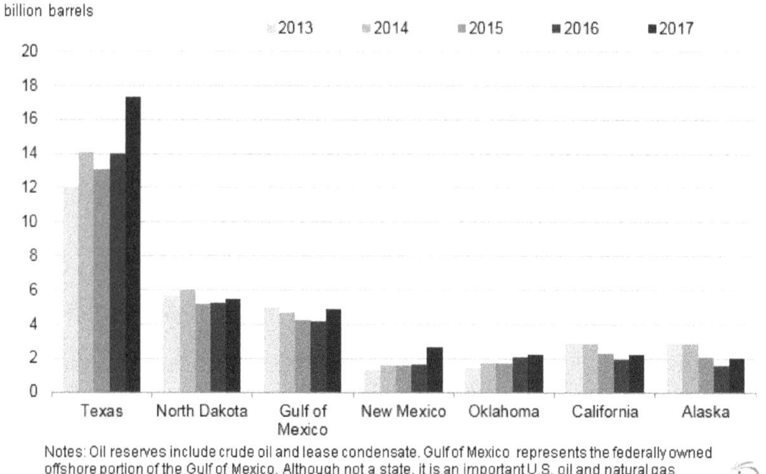

Notes: Oil reserves include crude oil and lease condensate. Gulf of Mexico represents the federally owned offshore portion of the Gulf of Mexico. Although not a state, it is an important U.S. oil and natural gas production area.
Source: U.S. Energy Information Administration, Form EIA-23L, *Annual Report of Domestic Oil and Gas Reserves*, 2013–17

## Offshore Reserves

The Minerals Management Service (MMS) estimates the Federal Outer Continental Shelf (OCS) contains between 66.6 and 115.1 billion barrels of undiscovered technically recoverable crude oil, with a mean estimate of 85.9 billion barrels. The Gulf of Mexico OCS ranks first with a mean estimate of 44.9 billion barrels, followed by Alaska OCS with 38.8 billion barrels. At $80/bbl crude prices, the MMS estimates that 70 billion barrels are economically recoverable. In 2008, a total of about 574 million acres of the OCS are off-limits to leasing and development. The moratoria and presidential withdrawal cover about 85 percent of OCS area offshore the lower 48 states. The MMS estimates that the resources in OCS areas currently off limits to leasing and development total 17.8 billion barrels.

## Arctic

In 1998, the USGS estimated that the 1002 area of the Arctic National Wildlife Refuge contains a total of between 5.7 and 16.0 billion barrels of undiscovered, technically recoverable oil, with a mean estimate of 10.4 billion barrels, of which 7.7 billion barrels falls within the Federal portion of the ANWR 1002 Area. In May 2008 the EIA used this assessment to estimate the potential cumulative production of the 1002 area of ANWR to be a maximum of 4.3 billion barrels from 2018 to 2030. This estimate is a best-case scenario of technically recoverable oil during the area's primary production years if legislation were passed in 2008 to allow drilling.

A 2002 assessment concluded that the National Petroleum Reserve–Alaska contains between 6.7 and 15.0 billion barrels of oil, with a mean (expected) value of 10.6 billion barrels. The quantity of undiscovered oil beneath Federal lands (excluding State and Native areas) is estimated to range between 5.9 and 13.2 BBO, with a mean value of 9.3 BBO. Most oil accumulations are expected to be of moderate size, on the order of 30 to 250 million barrels each. Large accumulations like the Prudhoe Bay oil field (whose ultimate recovery is approximately 13 billion barrels, are not expected to occur. The volumes of undiscovered, technically recoverable oil estimated for NPRA are similar to the volumes estimated for ANWR. However, because of differences in accumulation sizes (the ANWR study area is estimated to contain more accumulations in larger size classes) and differences in assessment area (the NPRA study area is more than 12 times larger than the ANWR study area), economically recoverable resources are different at low oil prices. But at market prices above $40 per barrel, estimates of economically recoverable oil for NPRA are similar to ANWR.

## Tight Oil

In April 2008, the USGS released a report giving a new resource assessment of the Bakken Formation underlying portions of Montana and North Dakota. The USGS believes that with new horizontal drilling technology there is somewhere between 3.0 and 4.5 billion barrels of undiscovered, technically recoverable oil in this 200,000 square miles formation that was initially discovered in 1951. If accurate, this reassessment would make it the largest "continuous" oil accumulation (The USGS uses "continuous" to describe accumulations requiring extensive artificial fracturing to allow the oil to flow to the borehole) ever discovered in the U.S. The formation is estimated to contain significantly more - figures in excess of 150 billion barrels have been reported - but it is yet

67

uncertain how much of this oil is recoverable using current technology. In 2011, Harold Hamm claimed that the recoverable share may reach 24 billion barrels; this would mean that Bakken contains more extractable petroleum than all other known oil fields in the country, combined.

## UNCONVENTIONAL PROSPECTIVE RESOURCES

### Oil Shale

The United States has the largest known deposits of oil shale in the world, according to the Bureau of Land Management and holds an estimated 2.175 trillion barrels of potentially recoverable oil. Oil shale does not actually contain oil, but a waxy oil precursor known as kerogen. There is no significant commercial production of oil from oil shale in the United States.

### Oil Sands

There are significant volumes of heavy oil in the oil sands of northeast Utah. There has yet to be any significant production from these deposits.

## OTHER CLASSIFICATIONS

A more sophisticated system of evaluating petroleum accumulations was adopted in 2007 by the Society of Petroleum Engineers (SPE), World Petroleum Council (WPC), American Association of Petroleum Geologists (AAPG), and Society of Petroleum Evaluation Engineers (SPEE). It incorporates the 1997 definitions for reserves, but adds categories for contingent resources and prospective resources.

Contingent resources are those quantities of petroleum estimated, as of a given date, to be potentially recoverable from known accumulations, but the applied project(s) are not yet considered mature enough for commercial development due to one or more contingencies. Contingent resources may include, for example, projects for which there are no viable markets, or where commercial recovery is dependent on technology under development, or where evaluation of the accumulation is insufficient to clearly assess commerciality.

Prospective resources are those quantities of petroleum estimated, as of a given date, to be potentially recoverable from undiscovered accumulations by application of future development projects. Prospective resources have both an associated chance of discovery and a chance of development.

The United States Geological Survey uses the terms technically and economically recoverable resources when making its petroleum resource assessments. Technically recoverable resources represent that proportion of assessed in-place petroleum that may be recoverable using current recovery technology, without regard to cost. Economically recoverable resources are technically recoverable petroleum for which the costs of discovery, development, production, and transport, including a return to capital, can be recovered at a given market price.

"Unconventional resources" exist in petroleum accumulations that are pervasive throughout a large area. Examples include extra heavy oil, oil sand, and oil shale deposits. Unlike "conventional resources", in which the petroleum is recovered through wellbores and typically requires minimal processing prior to sale, unconventional resources require specialized extraction technology to produce. For example, steam and/or solvents are used to mobilize bitumen for in-situ recovery. Moreover, the extracted petroleum may require significant processing prior to sale (e.g., bitumen upgraders). The total amount of unconventional oil resources in the world considerably exceeds the amount of conventional oil reserves, but are much more difficult and expensive to develop.

According to the EIA, the domestic production of energy from crude oil and distillates will continue to increase and will still be at above 2018 levels in 2050. In the following graph a quad is a unit of energy equal to 1015 (a short-scale quadrillion) BTU, or 1.055 × 1018 joules (1.055 exajoules or EJ) in SI units.

One quad = Approximately 8,000,000,000 gallons (US) of gasoline or 293,000,000,000 kilowatt-hours (kWh)

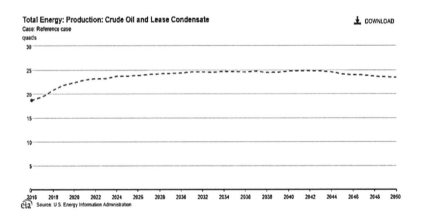

**CONCLUSION: CRUDE OIL WILL BE A SUBSTANTIAL COMPONENT OF THE ENERGY MIX WELL BEYOND 2050.**

## NATURAL GAS

Natural gas in the United States was the nation's largest source of energy production in 2016, representing 33 percent of all energy produced in the country. Natural gas has been the largest source of electrical generation in the United States since July 2015. Proved reserves of natural gas increased by 123.2 trillion cubic feet (Tcf) (36.1%) to 464.3 Tcf at year-end 2017—a new U.S. record for total natural gas proved reserves. The previous U.S. record was 388.8 Tcf, set in 2014. U.S. production of total natural gas increased by 4% from 2016 to 2017, reaching a new record level. The share of natural gas from shale compared with total U.S. natural gas proved reserves increased from 62% in 2016 to 66% at year-end 2017.

Proved natural gas reserves increased in each of the top eight U.S. natural gas reserves states in 2017 (Figure 3). Pennsylvania had the largest net increase in proved natural gas reserves of any state, adding 28.1 Tcf of proved natural gas reserves in the Marcellus and Utica shale plays. EQT Corporation announced on December 13, 2017, that it had successfully completed the longest lateral by any operator in the Marcellus shale play—the Haywood H18 well in Washington County, Pennsylvania, has a completed lateral length of 17,400 feet (more than three miles)—and that the company plans to drill 27 Marcellus wells at 17,000 feet or longer in 2018.

Texas had the second-largest net increase, adding 26.9 Tcf of proved natural gas reserves—the largest portion of the increase came from associated-dissolved natural gas proved reserves that accompanied the gains in crude oil proved reserves in the Permian Basin. The third-largest net increase in proved natural gas reserves was in Louisiana, where operators added 18.4 Tcf of proved reserves developing the Haynesville/Bossier shale play. (EIA, 2019)

Figure 3. Proved reserves of the top eight U.S. natural gas reserves states, 2013–17

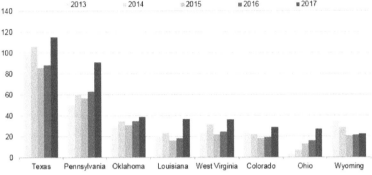

Note: Total natural gas includes natural gas plant liquids that have yet to be extracted downstream and does not include lease condensate.
Source: U.S. Energy Information Administration, Form EIA-23L, *Annual Report of Domestic Oil and Gas Reserves*, 2013–17

**CONCLUSION: NATURAL GAS WILL BE A SUBSTANTIAL COMPONENT OF THE ENERGY MIX WELL BEYOND 2050.**

## CRUDE OIL AND NATURAL GAS COMBINED

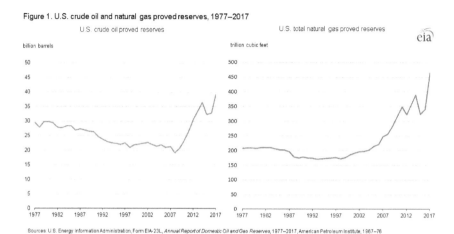

Figure 1. U.S. crude oil and natural gas proved reserves, 1977–2017

Sources: U.S. Energy Information Administration, Form EIA-23L, *Annual Report of Domestic Oil and Gas Reserves*, 1977–2017, American Petroleum Institute, 1967–76

These are the very energy resources that the Green New Deal proposes to completely remove from the U. S. Energy supply in 10 to 12 years. These carbon-based fuels; crude oil, natural gas and coal are currently providing 93 % of our total electrical energy requirements. They are reliable and constant. Technical improvements are continuing to reduce harmful emissions as well as those emissions that are less harmful but have, nonetheless, been deemed harmful by environmental activists (carbon dioxide).

### BIOMASS AND BIOFUELS

Biomass is waste material from plants or animals that is not used for food or feed; it can be waste from farming (like wheat stalks) or horticulture (yard waste), food processing (like corn cobs), animal farming (manure rich in nitrogen and phosphorus), or human waste from sewage plants. It is used in various industrial processes, like energy production or as raw materials for manufacturing chemicals.

Burning biomass releases carbon emissions, but has been classed as a renewable energy source in the EU and UN legal frameworks, because plant stocks can be replaced with new growth. It has become popular among coal power stations, which switch from coal to biomass in order to convert to renewable energy generation without wasting existing generating plant and infrastructure. Biomass most often refers to an energy source; biomass can either be used directly via combustion to produce heat, or indirectly after converting it to various forms of biofuel. Conversion of biomass to biofuel can be achieved by different methods which are broadly classified into: thermal, chemical, and biochemical. Some chemical constituents of plant biomass include lignins, cellulose, and hemicellulose. (Wkipedia, 2019)

In terms of how biomass is used as fuel, percentages were gathered in the United States of 2016. 5% is considered primary energy in the U.S. Making up that 5% of primary energy, about 48% comes from biofuels (mostly ethanol), 41% from wood-based biomass, and around 11% from municipal waste. Historically, humans have harnessed biomass-derived energy since the time when people began burning wood to make fire. Even in 2018, biomass is the only source of fuel for domestic use in many developing countries. Biomass is all biologically-produced matter based in carbon, hydrogen and oxygen. Wood remains the largest biomass energy source today; examples include forest residues (such as dead trees, branches and tree stumps), yard clippings, wood chips and even municipal solid waste. Wood energy is derived by using lignocellulosic biomass (second-generation biofuels) as fuel. Harvested wood may be used directly as a fuel or collected from wood waste streams to be processed into pellet fuel or other forms of fuels.

Biomass also includes plant or animal matter that can be converted into fibers or other industrial chemicals, including biofuels. Industrial biomass can be grown from numerous types of plants, including miscanthus, switchgrass, hemp, corn, poplar, willow, sorghum, sugarcane, bamboo, and a variety of tree species, ranging from eucalyptus to oil palm (palm oil).

Based on the source of biomass, biofuels are classified broadly into two major categories. First-generation biofuels are derived from sources such as sugarcane and corn starch. Sugars present in this biomass are fermented to produce bioethanol, an alcohol fuel which can be used directly in a fuel cell to produce electricity or serve as an additive to gasoline. However, utilizing food-based resources for fuel production only aggravates the food shortage problem. Second-generation biofuels, on the other hand, utilize non-food-based biomass sources such as agriculture and municipal waste. These biofuels mostly consist of lignocellulosic biomass, which is not edible and is a low-value waste for many industries. Despite being the favored alternative, economical production of second-generation biofuel is not yet achieved due to technological issues. These issues arise mainly due to chemical inertness and structural rigidity of lignocellulosic biomass.

Plant energy is produced by crops specifically grown for use as fuel that offer high biomass output per hectare with low input energy. Some examples of these plants are wheat, which typically yields 7.5–8 metric tons of grain per hectare [one hectare contains about 2.47 acres], and straw, which typically yields 3.5–5 metric tons per hectare. The grain can be used for liquid transportation fuels while the straw can be burned to produce heat or electricity. Plant biomass can also be degraded from cellulose to glucose through a series of chemical treatments, and the resulting sugar can then be used as a first-generation biofuel.

The main contributors of waste energy are municipal solid waste, manufacturing waste, and landfill gas. Energy derived from biomass is projected to be the largest non-hydroelectric renewable resource of electricity in the US between 2000 and 2020.

Biomass can be converted to other usable forms of energy like methane gas or transportation fuels like ethanol and biodiesel. Rotting garbage, and agricultural and human waste, all release methane gas, also called landfill gas or biogas. Crops such as corn and sugarcane can be fermented to produce the transportation fuel ethanol. Biodiesel, another transportation fuel, can be produced from leftover food products like vegetable oils and animal fats. Several biodiesel companies simply collect used restaurant cooking oil and convert it into biodiesel. Also, biomass-to-liquids (called "BTLs") and cellulosic ethanol are still under research.

There is research involving algae or algae-derived biomass, as this non-food resource can be produced at rates five to ten times those of other types of land-based agriculture, such as corn and soy. Once harvested, it can be fermented to produce biofuels such as ethanol, butanol, and methane, as well as biodiesel and hydrogen. Efforts are being made to identify which species of algae are most suitable for energy production. Genetic engineering approaches could also be utilized to improve microalgae as a source of biofuel.

The biomass used for electricity generation varies by region. Forest by-products, such as wood residues, are common in the US. Agricultural waste is common in Mauritius (sugar cane residue) and Southeast Asia (rice husks).

Sewage sludge can be another source of biomass. For example, the Omni Processor is a process which uses sewage sludge as fuel in a process of sewage sludge treatment, with surplus electrical energy being generated for export.

Using biomass as a fuel produces air pollution in the form of carbon monoxide, carbon dioxide, NOx (nitrogen oxides), VOCs (volatile organic compounds), particulates and other pollutants at levels above those from traditional fuel sources such as coal or natural gas in some cases (such as with indoor heating and cooking). Use of wood biomass as a fuel can also produce fewer particulate and other pollutants than open burning as seen in wildfires or direct heat applications.

The biomass power generating industry in the United States consists of approximately 11,000 MW of summer operating capacity actively supplying power to the grid, and produces about 1.4 percent of the U.S. electricity supply. (EIA, 2019)

## LOW CARBON CONTENT ENERGY RESOURCES
### NUCLEAR FISSION

Nuclear power plants use heat produced during nuclear fission to heat water to produce steam. The steam is used to spin large turbines that generate electricity. In nuclear fission, atoms are split apart to form smaller atoms, releasing energy. Fission takes place inside the reactor of a nuclear power plant. At the center of the reactor is the core, which contains uranium fuel.

The uranium fuel is formed into ceramic pellets. Each ceramic pellet produces about the same amount of energy as 150 gallons of oil. These energy-rich pellets are stacked end-to-end in 12-foot metal fuel rods. A bundle of fuel rods, some with hundreds of rods, is called a fuel assembly. A reactor core contains many fuel assemblies.

The heat produced during nuclear fission in the reactor core is used to boil water into steam, which turns the blades of a steam turbine. As the turbine blades turn, they drive generators that make electricity. Nuclear plants cool the steam back into water in a separate structure at the power plant called a cooling tower or they use water from ponds, rivers, or the ocean. The cooled water is then reused to produce steam. Nuclear power plants generate about 20% of U.S. electricity. As of January 2018, there were 99 operating nuclear reactors at 61 nuclear power plants in 30 states. Thirty-six of the plants have 2 or more reactors.

The United States generates more nuclear power than any other country. Of the 31 countries in the world with commercial nuclear power plants in 2015, the United States had the most nuclear electricity generation capacity and generated more electricity from nuclear energy than any other country. France had the second-highest nuclear electricity generation capacity and electricity generation and obtained about 78% of its total electricity generation from nuclear energy, the largest share of any other country. Fourteen other countries generated at least 20% of their electricity from nuclear power.

Nuclear reactors in the United States may have large concrete domes covering the reactors, which are required to contain accidental releases of radiation. Not all nuclear power plants have cooling towers. Some nuclear power plants use water from lakes, rivers, or the ocean for cooling. (EIA on Nuclear Power Plants, 2019)

The actual fuel used by a nuclear power plant has discussed in the preceding is uranium. Any discussion or plan on Energy must consider the actual fuel, not just the power generation process. However, converting uranium to fuel also involves a more complex and expensive process than converting crude oil and natural gas to fuel.

Uranium is the fuel most widely used by nuclear plants for nuclear fission. Uranium is considered a nonrenewable energy source, even though it is a common metal found in rocks worldwide. Nuclear power plants use a certain kind of uranium, referred to as U-235, for fuel because its atoms are easily split apart. Although uranium is about 100 times more common than silver, U-235 is relatively rare.

Most U.S. uranium ore is mined in the western United States. Once uranium is mined, the U-235 must be extracted and processed before it can be used as a fuel.

The nuclear fuel cycle consists of front-end steps that prepare uranium for use in nuclear reactors and back-end steps to safely manage, prepare, and dispose of highly radioactive spent nuclear fuel. Chemical processing of spent fuel material to recover any remaining product that could undergo fission again in a new fuel assembly is technically feasible, but it is not permitted in the United States.

In much the same manner as the exploration for coal, oil and natural gas, the nuclear fuel cycle starts with exploration for uranium and the development of mines to extract uranium ore. A variety of techniques are used to locate uranium, such as airborne radiometric surveys, chemical sampling of groundwater and soils, and exploratory drilling to understand the underlying geology. Once uranium ore deposits are located, the mine developer usually follows up with more closely spaced in fill, or development drilling, to determine how much uranium is available and what it might cost to recover it.

When ore deposits that are economically feasible to recover are located, the next step in the fuel cycle is to mine the ore using one of the following techniques: underground mining, open pit mining, in-place (in-situ) solution mining, and heap leaching.

Before 1980, most U.S. uranium was produced using open pit and underground mining techniques. Today, most U.S. uranium is produced using a solution mining technique commonly called in-situ-leach (ISL) or in-situ-recovery (ISR). This process extracts uranium that coats the sand and gravel particles of groundwater reservoirs. The sand and gravel particles are exposed to a solution with a pH that has been elevated slightly by using oxygen, carbon dioxide, or caustic soda. The uranium dissolves into the groundwater, which is pumped out of the reservoir and processed at a uranium mill. Heap leaching involves spraying an acidic liquid solution onto piles of crushed uranium ore. The solution drains down through the crushed ore and leaches uranium out of the rock, which is recovered from underneath the pile. Heap leaching is no longer used in the United States.

After the uranium ore is extracted from an open pit or underground mine, it is refined into uranium concentrate at a uranium mill. The ore is crushed, pulverized, and ground into a fine powder. Chemicals are added to the fine powder, which causes a reaction that separates the uranium from the other minerals. Groundwater from solution mining operations is circulated through a resin bed to extract and concentrate the uranium.

The concentrated uranium product is typically a bright yellow or orange powder called yellowcake ($U_3O_8$). The solid waste material from pit and underground mining operations is called mill tailings. The processed water from solution mining is returned to the groundwater reservoir where the mining process is repeated.

The next step in the nuclear fuel cycle is to convert yellowcake into uranium hexafluoride ($UF_6$) gas at a converter facility. Three forms (isotopes) of uranium occur in nature: U-234, U-235, and U-238. Current U.S. nuclear reactor designs require a stronger concentration (enrichment) of the U-235 isotope to operate efficiently. The uranium hexafluoride gas produced in the converter facility is called natural $UF_6$ because the original concentrations of uranium isotopes are unchanged.

After conversion, the $UF_6$ gas is sent to an enrichment plant where the individual uranium isotopes are separated to produce enriched $UF_6$, which has a 4% to 5% concentration of U-235.

Two types of uranium enrichment processes have been used in the United States: gaseous diffusion and gas centrifuge. The United States currently has one operating enrichment plant, which uses a gas centrifuge process. Enriched $UF_6$ is sealed in canisters and allowed to cool and solidify before it is transported to a nuclear reactor fuel assembly plant by train, truck, or barge.

Atomic vapor laser isotope separation (AVLIS) and molecular laser isotope separation (MLIS) are new enrichment technologies currently under development. These laser-based enrichment processes can achieve higher initial enrichment (isotope separation) factors than the diffusion or centrifuge processes and can produce enriched uranium more quickly than other techniques.

Once the uranium is enriched, it is ready to be converted into nuclear fuel. At a nuclear fuel fabrication facility, the UF6, in solid form, is heated to gaseous form, and then the UF6 gas is chemically processed to form uranium dioxide (UO2) powder. The powder is then compressed and formed into small ceramic fuel pellets. The pellets are stacked and sealed into long metal tubes that are about 1 centimeter in diameter to form fuel rods. The fuel rods are then bundled together to make up a fuel assembly. Depending on the reactor type, each fuel assembly has about 179 to 264 fuel rods. A typical reactor core holds 121 to 193 fuel assemblies.

Once the fuel assemblies are fabricated, trucks transport them to the reactor sites. The fuel assemblies are stored onsite in fresh fuel storage bins until the reactor operators need them. At this stage, the uranium is only mildly radioactive, and essentially all radiation is contained within the metal tubes. Typically, reactor operators change out about one-third of the reactor core (40 to 90 fuel assemblies) every 12 to 24 months.

The reactor core is a cylindrical arrangement of the fuel bundles that is about 12 feet in diameter and 14 feet tall and encased in a steel pressure vessel with walls that are several inches thick. The reactor core has essentially no moving parts except for a small number of control rods that are inserted to regulate the nuclear fission reaction. Placing the fuel assemblies next to each other and adding water initiates the nuclear reaction.

After use in the reactor, fuel assemblies become highly radioactive and must be removed and stored under water at the reactor site in a spent fuel pool for several years. Even though the fission reaction has stopped, the spent fuel continues to give off heat from the decay of the radioactive elements that were created when the uranium atoms were split apart. The water in the pool serves to both cool the fuel and block the release of radiation. From 1968 through June 2013, 241,468 fuel assemblies had been discharged and stored at 118 commercial nuclear reactors in the United States.

Within a few years, the spent fuel cools in the pool and may be moved to a dry cask storage container at the power plant site. An increasing number of reactor operators now store their older spent fuel in these special outdoor concrete or steel containers with air cooling.

The final step in the nuclear fuel cycle is the collection of spent fuel assemblies from the interim storage sites for final disposition in a permanent underground repository. The United States currently has no permanent underground repository for high-level nuclear waste.

In 2017, about 45.5 million pounds of uranium (U3O8 equivalent) were loaded into commercial U.S. nuclear power reactors. (EIA on Nuclear Fuel, 2019)

## HYDROELECTRIC POWER

Hydroelectric power or Hydropower is from energy in moving water. Humans have a long history of using the force of water flowing in streams and rivers to produce mechanical energy. Hydropower was one of the first sources of energy used for electricity generation and is the largest single renewable energy source for electricity generation in the United States.

In 2017, hydroelectricity accounted for about 7.5% of total U.S. utility-scale electricity generation and 44% of total utility-scale electricity generation from renewable energy sources. Hydroelectricity's share of total U.S. electricity generation has decreased over time, mainly because electricity generation from other sources has increased.

Hydropower relies on the water cycle. Understanding the water cycle is important to understanding hydropower. The water cycle has three steps:

-Solar energy heats water on the surface of rivers, lakes, and oceans, which causes the water to evaporate.

-Water vapor condenses into clouds and falls as precipitation—rain and snow.

-Precipitation collects in streams and rivers, which empty into oceans and lakes, where it evaporates and begins the cycle again.

The amount of precipitation that drains into rivers and streams in a geographic area determines the amount of water available for producing hydropower. Seasonal variations in precipitation and long-term changes in precipitation patterns, such as droughts, have a big impact on hydropower production.

Hydroelectric power is produced from moving water. Because the source of hydroelectric power is water, hydroelectric power plants are usually located on or near a water source. The volume of the water flow and the change in elevation (or fall) from one point to another determine the amount of available energy in moving water. Swiftly flowing water in a big river, such as the Columbia River that forms the border between Oregon and Washington, carries a great deal of energy in its flow. Water descending rapidly from a high point, such as Niagara Falls in New York, also has substantial energy in its flow.

At both Niagara Falls and the Columbia River, water flows through a pipe, or penstock, then pushes against and turns blades in a turbine to spin a generator to produce electricity. In a run-of-the-river system, the force of the current applies pressure on a turbine. In a storage system, water accumulates in reservoirs created by dams and is released as needed to generate electricity.

The first U.S. hydroelectric power plant opened on the Fox River near Appleton, Wisconsin, on September 30, 1882. Most of U.S. hydroelectricity is now produced at large dams on major rivers, and most of these hydroelectric dams were built before the mid-1970's.

The gravitational pull of the moon and sun along with the rotation of the earth cause the tides. In some places, tides cause water levels near the shore to vary up to 40 feet. People in Europe harnessed this movement of water to operate grain mills more than a 1,000 years ago. Today, tidal energy systems generate electricity. Producing tidal energy economically requires a tidal range of at least 10 feet.

### Tidal Power Barrage

One type of tidal energy system uses a structure similar to a dam called a barrage. The barrage is installed across an inlet of an ocean bay or lagoon that forms a tidal basin. Sluice gates on the barrage control water levels and flow rates to allow the tidal basin to fill on the incoming high tides and to empty through an electricity turbine system on the outgoing ebb tide. A two-way tidal power system generates electricity from both the incoming and outgoing tides.

A potential disadvantage of tidal power is the effect a tidal station can have on plants and animals in estuaries of the tidal basin. Tidal barrages can change the tidal level in the basin and increase turbidity (the amount of matter in suspension in the water). They can also affect navigation and recreation.

Several tidal power barrages operate around the world; however, the United States does not have any tidal power plants, and it only has a few sites where tidal energy could be economical to produce. France,

England, Canada, and Russia have much more potential to use tidal power.

## Tidal Turbines

Tidal turbines look similar to wind turbines. They can be placed on the sea floor where there is strong tidal flow. Because water is about 800 times denser than air, tidal turbines have to be much sturdier and heavier than wind turbines. Tidal turbines are more expensive to build than wind turbines but capture more energy with the same size blades. The tidal turbine projects in Scotland and South Korea have tidal turbines with 1.5 MW electricity generation capacity. The project in Scotland is planning to have up to 400 MW of electricity generation capacity. A demonstration tidal turbine project is under development in the East River of New York.

## Tidal Fences

A tidal fence is a type of tidal power system that has vertical axis turbines mounted in a fence or row placed on the sea bed, similar to tidal turbines. Water passing through the turbines generates electricity. As of the end of 2017, no tidal fence projects were operating.

## Wave Energy

Waves form as wind blows over the surface of open water in oceans and lakes. Ocean waves contain tremendous energy. The theoretical annual energy potential of waves off the coasts of the United States is estimated to be as much as 2.64 trillion kilowatt-hours, or the equivalent of about 66% of U.S. electricity generation in 2017. The west coasts of the United States and Europe, and the coasts of Japan and New Zealand, have potential sites for harnessing wave energy.

There are various ways to channel the power of waves. One way to harness wave energy is to bend or focus waves into a narrow channel to increase their size and power and to spin the turbines that generate electricity. Waves can also be channeled into a catch basin or reservoir where the water flows to a turbine at a lower elevation, similar to the way a hydropower dam operates.

Many other methods of capturing wave energy are under development. These methods include placing devices on or just below the surface of the water and anchoring devices to the ocean floor. The U.S. Department of Energy's Marine and Hydrokinetic Technology Database provides information on marine and hydrokinetic renewable energy, both in the U.S. and around the world.

## Ocean Thermal Energy Conversion

Ocean thermal energy conversion (OTEC) is a process or technology for producing energy by harnessing the temperature differences (thermal gradients) between ocean surface waters and deep ocean waters.

Energy from the sun heats the surface water of the ocean. In tropical regions, surface water can be much warmer than deep water. This temperature difference can be used to produce electricity and to desalinate ocean water. Ocean Thermal Energy Conversion (OTEC) systems use a temperature difference (of at least 77o Fahrenheit) to power a turbine to produce electricity. Warm surface water is pumped through an evaporator containing a working fluid. The vaporized fluid drives a turbine/generator. The vaporized fluid is turned back to a liquid in a condenser cooled with cold ocean water pumped from deeper in the ocean. OTEC systems using seawater as the working fluid can use the condensed water to produce desalinated water.

The United States became involved in OTEC research in 1974 with the establishment of the Natural Energy Laboratory of Hawaii Authority. The laboratory is one of the world's leading test facilities for OTEC technology. The laboratory operated a 250 kilowatt (kW) demonstration OTEC plant for six years in the 1990s. The United States Navy supported the development of a 105 kW demonstration OTEC plant at the laboratory site. This facility became operational in 2015 and supplies electricity to the local electricity grid.

Other larger OTEC systems are in development or planned in several countries, mostly to supply electricity and desalinated water for island communities. (EIA, 2019)

### GEOTHERMAL

Geothermal energy is heat within the earth. The word geothermal comes from the Greek words geo (earth) and therme (heat). Geothermal energy is a renewable energy source because heat is continuously produced inside the earth. People use geothermal heat for bathing, to heat buildings, and to generate electricity.

The slow decay of radioactive particles in the earth's core, a process that happens in all rocks, produces geothermal energy. The earth has four major parts or layers: An inner core of solid iron that is about 1,500 miles in diameter, an outer core of hot molten rock called magma that is about 1,500 miles thick, a mantle of magma and rock surrounding the outer core that is about 1,800 miles thick and, a crust of solid rock that forms the continents and ocean floors that is 15 to 35 miles thick under the continents and 3 to 5 miles thick under the oceans. Scientists have discovered that the temperature of the earth's inner core is about 10,800 degrees Fahrenheit (°F), which is as hot as the surface of the sun. Temperatures in the mantle range from about 392°F at the upper boundary with the earth's crust to approximately 7,230°F at the mantle-core boundary.

The earth's crust is broken into pieces called tectonic plates. Magma comes close to the earth's surface near the edges of these plates, which is where many volcanoes occur. The lava that erupts from volcanoes is partly magma. Rocks and water absorb heat from magma deep underground. The rocks and water found deeper underground have the highest temperatures.

Geothermal reservoirs are naturally occurring areas of hydrothermal resources. These reservoirs are deep underground and are largely undetectable above ground. Geothermal energy finds its way to the earth's surface in three ways: Volcanoes and fumaroles (holes in the earth where volcanic gases are released), Hot springs and Geysers.

Most geothermal resources are near the boundaries of the earth's tectonic plates. The most active geothermal resources are usually found along major tectonic plate boundaries where most volcanoes are located. One of the most active geothermal areas in the world is called the Ring of Fire, which encircles the Pacific Ocean.

Most of the geothermal power plants in the United States are in western states and Hawaii, where geothermal energy resources are close to the earth's surface. California generates the most electricity from geothermal energy. The Geysers dry steam reservoir in Northern California is the largest known dry steam field in the world and has been producing electricity since 1960.

Geothermal power plants use hydrothermal resources that have both water (hydro) and heat (thermal). Geothermal power plants require high-temperature (300°F to 700°F) hydrothermal resources that come from either dry steam wells or from hot water wells. People use these resources by drilling wells into the earth and then piping steam or hot water to the surface. The hot water or steam powers a turbine that generates electricity. Some geothermal wells are as much as two miles deep.

There are three basic types of geothermal power plants: Dry steam plants use steam directly from a geothermal reservoir to turn generator turbines, flash steam plants that take high-pressure hot water from deep inside the earth and convert it to steam to drive generator turbines and binary cycle power plants that transfer the heat from geothermal hot water to another liquid. The heat causes the second liquid to turn to steam, which is used to drive a generator turbine. (EIA, 2019)

### HYDROGEN: The Ultimate Clean Fuel

The sun is essentially a giant ball of hydrogen gas undergoing fusion into helium gas. This process causes the sun to produce vast amounts of energy. Hydrogen is the lightest element. Hydrogen is a gas at normal temperature and pressure, but hydrogen condenses to a liquid at -423° Fahrenheit (-253° Celsius). Hydrogen is the simplest element. Each atom of hydrogen has only one proton. Hydrogen is also the most plentiful gas in the universe. Stars like the sun consist primarily of hydrogen.

Hydrogen occurs naturally on earth only in compound form with other elements in liquids, gases, or solids. Hydrogen combined with oxygen is water ($H_2O$). Hydrogen combined with carbon forms different compounds—or hydrocarbons—found in natural gas, coal, and petroleum. Hydrogen has the highest energy content of any common fuel by weight (about three times more than gasoline), but it has the lowest energy content by volume (about four times less than gasoline).

Hydrogen is an energy carrier. Energy carriers allow the transport of energy in a useable form from one place to another. Hydrogen, like electricity, is an energy carrier that must be produced from another substance. Hydrogen can be produced—separated—from a variety of sources including water, fossil fuels, or biomass, and used as a source of energy or fuel.

It takes more energy to produce hydrogen (by separating it from other elements in molecules) than the hydrogen provides when it is converted to useful energy. However, hydrogen is useful as an energy source/fuel because it has a high energy content per unit of weight, which is why it is used as a rocket fuel and in fuel cells to produce electricity on some spacecraft. Hydrogen is not widely used as a fuel for other uses now, but it has the potential for greater use in the future.

To produce hydrogen, it must be separated from the other elements in the molecules where it occurs. Hydrogen atoms can be separated from water, hydrocarbons in coal, petroleum, and natural gas, and from biomass. The two most common methods for producing hydrogen are steam reforming and electrolysis (water splitting).

Steam reforming is currently the least expensive way to produce hydrogen, and it accounts for most of the commercially produced hydrogen in the United States. This method is used in industries to separate hydrogen atoms from carbon atoms in methane ($CH_4$). The steam reforming process results in carbon dioxide emissions.

Electrolysis is a process that splits hydrogen from water using an electric current. The process can be used on a large or small scale. Electrolysis does not produce any emissions other than hydrogen and oxygen, and the electricity used in electrolysis can come from renewable sources such as hydro, wind, or solar energy. If the electricity used in electrolysis is produced from fossil fuels, then the pollution and carbon dioxide emissions produced from those fuels are indirectly associated with electrolysis.

Research is underway to develop other ways to produce hydrogen. These methods include using microbes that use light to make hydrogen, converting biomass into liquids and separating the hydrogen, and using solar energy technologies to split hydrogen from water molecules.

Nearly all of the hydrogen consumed in the United States is used by industry for refining petroleum, treating metals, producing fertilizer, and processing foods. Rocket fuel is the main use of hydrogen for energy. The National Aeronautics and Space Administration (NASA) is the largest user of hydrogen as a fuel. NASA began using liquid hydrogen in the 1950s as a rocket fuel, and NASA was one of the first to use fuel cells to power the electrical systems on space craft.

Hydrogen fuel cells produce electricity by combining hydrogen and oxygen atoms. This combination results in an electrical current. A fuel cell is two to three times more efficient than an internal combustion engine running on gasoline.

Many different types of fuel cells are available for a wide range of applications. Small fuel cells can power laptop computers, cell phones, and military applications. Large fuel cells can provide electricity for emergency power in buildings and in remote areas that are not connected to electric power grids. Hydrogen use in vehicles is a major focus of fuel cell research and development.

The interest in hydrogen as an alternative transportation fuel is based on its potential for domestic production, its use in fuel cells for zero-emission electric vehicles, and the fuel cell vehicle's potential for high efficiency. In the United States, several vehicle manufacturers have begun making light-duty hydrogen fuel cell electric vehicles available in select regions like southern and northern California where there is access to hydrogen fueling stations. Test vehicles are also available in limited numbers to select organizations with access to hydrogen fueling stations.

Most hydrogen-fueled vehicles are automobiles and transit buses that have an electric motor powered by a fuel cell. A few of these vehicles burn hydrogen directly. The high cost of fuel cells and the limited availability of hydrogen fueling stations have limited the number of hydrogen-fueled vehicles.

Production of hydrogen-fueled cars is limited because people won't buy those cars if hydrogen refueling stations are not easily accessible, and companies won't build refueling stations if they don't have customers with hydrogen-fueled vehicles. In the United States, about 60 hydrogen refueling stations for vehicles are operating. About 40 of these stations are available for public use, nearly all of which are in California. The State of California has a program to help fund the development of publicly accessible hydrogen refueling stations throughout California to promote a consumer market for zero-emission fuel cell vehicles. (EIA, 2019)

### NUCLEAR FUSION-The Ultimate Energy Source?

Fusion power is a theoretical form of power generation in which energy will be generated by using nuclear fusion reactions to produce heat for electricity generation. In a fusion process, two lighter atomic nuclei combine to form a heavier nucleus, and at the same time, they release energy. This is the same process that powers stars like our Sun. Devices designed to harness this energy are known as fusion reactors.

Fusion processes require fuel and a highly confined environment with a high temperature and pressure, to create a plasma in which fusion can occur. In stars, the most common fuel is hydrogen, and gravity creates the high temperature and confinement needed for fusion. Fusion reactors generally use hydrogen isotopes such as deuterium and tritium, which react more easily, and create a confined plasma of millions of degrees using inertial methods (laser) or magnetic methods (tokamak and similar), although many other concepts have been attempted. The major challenges in realizing fusion power are to engineer a system that can confine the plasma long enough at high enough temperature and density for a long-term reaction to occur and, for the most common reactions, managing neutrons that are released during the reaction, which over time can degrade many common materials used within the reaction chamber.

As a source of power, nuclear fusion is expected to have several theoretical advantages over fission. These include reduced radioactivity in operation and little nuclear waste, ample fuel supplies, and increased safety. However, controlled fusion has proven to be extremely difficult to produce in a practical and economical manner. Research into fusion reactors began in the 1940s, but to date, no design has produced more fusion power output than the electrical power input; therefore, all existing designs have had a negative power balance.

Over the years, fusion researchers have investigated various confinement concepts. The early emphasis was on three main systems: z-pinch, stellarator and magnetic mirror. The current leading designs are the tokamak and inertial confinement (ICF) by laser. Both designs are being built at very large scales, most notably the ITER tokamak in France, and the National Ignition Facility laser in the United States. Researchers are also studying other designs that may offer cheaper approaches. Among these alternatives there is increasing interest in magnetized target fusion and inertial electrostatic confinement.

A December 29, 2017 NBC News report by Tom Metcalf included the following: " . . . Another form of nuclear energy known as fusion, which joins atoms of cheap and abundant hydrogen, can produce essentially limitless supplies of power without creating lots of radioactive waste.

Fusion has powered the sun for billions of years. Yet despite decades of effort, scientists and engineers have been unable to generate sustained nuclear fusion here on Earth. In fact, it's long been joked that fusion is 50 years away, and will always be.

But now it looks as if the long wait for commercial fusion power may be coming to an end — and sooner than in half a century . . .

. . . 'One of the brightest hopes for controlled nuclear fusion, the giant ITER reactor at Cadarache in southeastern France, is now on track to achieve nuclear fusion operation in the mid- to late-2040's', says Dr. William Madia, a former director of Oak Ridge National Laboratory who led an independent review of the ITER project in 2013. Madia says the decades needed to bring the ITER reactor to full operation reflect the huge engineering challenges still facing fusion researchers. These include building reactor walls that can withstand the intense heat of the fusion reaction — about 150 million degrees Celsius (270 million degrees Fahrenheit), or 10 times hotter than the core of the sun.

And then there's the challenge of creating superconducting materials that can generate the powerful magnetic fields needed to hold the fusion reaction in place. ITER has international backing and a budget of more than $14 billion. But it's not the only promising effort in the long quest for sustained nuclear fusion, or what some have called a "star in a jar."

Several smaller fusion projects, including commercial reactors being developed by Lockheed Martin in the U.S., General Fusion in Canada, and Tokamak Energy in the U.K., aim to feed fusion-generated power to electricity grids years before ITER produces its first fusion reactions.

'Our target is to deliver commercial power to the grid by 2030,' says Tokamak Energy's founder, Dr. David Kingham. The company foresees its fusion reactors replacing the fission reactors used in warships and submarines — and being put on trucks so they can be deployed wherever power is needed. A 100-megawatt fusion reactor that fits on the back of a truck could generate enough power for 100,000 people, according to the company.

Other fusion power projects include the Wendelstein 7-X fusion reactor in Germany, which uses an alternative to ITER's tokamak design known as a stellarator. Like ITER, the German reactor is backed by an international consortium and serves mainly for experimental research.

Exactly which, if any, of these initiatives will crack the fusion nut is still uncertain. But experts hope fusion power one day can make fossil-fuel-fired plants and nuclear fission reactors obsolete, along with most of their environmental problems.

And we can take heart that the remaining challenges are all just a matter of advanced engineering. Says Madia, 'We know the science is absolutely real because we can see it happening in the sun every day.'" (Metcalf, 2017)

## DISTRIBUTION

### THE POWER GRID

Electricity is generated at power plants and moves through a complex system, sometimes called the grid, of electricity substations, transformers, and power lines that connect electricity producers and consumers. Most local grids are interconnected for reliability and commercial purposes, forming larger, more dependable networks that enhance the coordination and planning of electricity supply.

At the beginning of the 20th century, more than 4,000 individual electric utilities operated in isolation from each other. As the demand for electricity grew, especially after World War II, utilities began to connect their transmission systems. These connections allowed utilities to share the economic benefits of building large and often jointly-owned electric generating units to serve their combined electricity demand at the lowest possible cost. Interconnection also reduced the amount of extra generating capacity that each utility had to hold to ensure reliable service during times of peak demand. Over time, three large, interconnected systems evolved in the United States.

In the United States, the entire electricity grid consists of hundreds of thousands of miles of high-voltage power lines and millions of miles of low-voltage power lines with distribution transformers that connect thousands of power plants to hundreds of millions of electricity customers all across the country.

## EXISTING

The stability of the electricity grid requires the electricity supply to constantly meet electricity demand, which in turn requires coordination of numerous entities that operate different components of the grid. The U.S. electricity grid consists of three large interconnected systems that operate to ensure its stability and reliability. To ensure coordination of electric system operations, the North American Electric Reliability Corporation developed and enforces mandatory grid reliability standards that the Federal Energy Regulatory Commission (FERC) approved. Local electricity grids are interconnected to form larger networks for reliability and commercial purposes. At the highest level, the U.S. power system in the Lower 48 states is made up of three main interconnections, which operate largely independently from each other with limited transfers of electricity between them. The Eastern Interconnection encompasses the area east of the Rocky Mountains and a portion of the Texas panhandle. The Western Interconnection encompasses the area from the Rockies to the west. The Electric Reliability Council of Texas (ERCOT) covers most of Texas.

## THE SMART GRID

The smart grid incorporates digital technology and advanced instrumentation into the traditional electrical system, which allows utilities and customers to receive information from and communicate with the grid. A smarter grid makes the electrical system more reliable and efficient by helping utilities reduce electricity losses and to detect and fix problems more quickly. The smart grid can help consumers intelligently manage energy use, especially at times when demand reaches significantly high levels or when a reduced energy demand is needed to support system reliability.

Smart devices in homes, offices, and factories can inform consumers and their energy management systems of times when an appliance is using relatively higher-priced electricity. These alerts help consumers, or their intelligent systems, to optimally adjust settings that, when supported by demand reduction incentives or time-of use electricity rates, can lower their energy bills. Smart devices on transmission and distribution lines and at substations allow a utility to more efficiently manage voltage levels and more easily find out where an outage or other problem is on the system. Smart grids can sometimes remotely correct problems in the electrical distribution system by digitally sending instructions to equipment that can adjust the conditions of the system. (EIA, 2019)

### STORAGE

The U.S. has about 23 gigawatts (GW) of storage capacity, approximately equal to the capacity of 38 typical coal plants.

Pumped hydroelectric storage accounts for about 96 percent of this total storage capacity, most of which was built in the 1960s and 1970s to accompany the new fleet of nuclear power plants. Because nuclear power plants are not designed to ramp up or down, their generation is constant at all times of the day. When demand for electricity is low at night, pumped hydro facilities store the energy from nuclear plants for later use during peak demand. These pumped hydro plants have proven valuable for quickly adjusting to small changes in demand or supply.

"Emerging storage facilities will allow us to store energy generated from wind and solar resources on shorter time frames to smooth variability, and on longer cycles to replace ever more fossil fuel. By charging storage facilities with energy generated from renewable sources, we can reduce our greenhouse gas emissions and our dependence on fossil fuels.

While the U.S. electric grid does not necessarily need more storage now, storage capacity will become more important as wind, solar, and other variable renewable energy resources expand in the power mix. Studies have shown that the existing grid can accommodate a sizeable increase in variable generation, but there are many exciting technologies in development that could help us store energy in the future and support an even greater amount of renewable energy on the grid.

Different energy storage technologies contribute to electricity stability by working at various stages of the grid, from generation to consumer end-use." (Union of Concerned Scientists (US), 2019)

Additional large-scale testing of hybrid electric systems has been proposed by Dr. John Deutch, an Institute Professor at the Massachusetts Institute of Technology. According to Dr. Deutch, "The declining cost of hybrid electric systems (HES) involving solar, wind, and energy storage suggests that carbon free (or nearly free) electricity generation is becoming competitive with combined cycle natural gas . . . . the best way to establish the credibility of the HES pathway is to organize an early large-scale demonstration of the operation and cost of a carbon free (or nearly free) HES ability to reliably meet the electricity demand of a relatively large service area rather than rely on government sponsored large scale demonstration projects or regulatory mandates compelling deployment of storage." (Deutch, 2019)

### Thermal Storage

Thermal storage is used for electricity generation by using power from the sun, even when the sun is not shining. Concentrating solar plants can capture heat from the sun and store the energy in water, molten salts, or other fluids. This stored energy is later used to generate electricity, enabling the use of solar energy even after sunset.

Plants like these are currently operating or proposed in California, Arizona, and Nevada. For example, the proposed Rice Solar Energy

Project in Blythe, California will use a molten salt storage system with a concentrating solar tower to provide power for approximately 68,000 homes each year.

Thermal storage technologies also exist for end-use energy storage. One method is freezing water at night using off-peak electricity, then releasing the stored cold energy from the ice to help with air conditioning during the day.

For example, Ice Energy's Ice Bear system creates a block of ice at night, and then uses the ice during the day to condense the air conditioning system's refrigerant. In this way, the Ice Bear system shifts the building's electricity consumption from the daytime peak to off-peak times when the electricity is less expensive. Additionally, the Bonneville Power Administration is conducting a pilot program on storing excess wind generation in residential water heaters.

### Compressed Air

Compressed Air Energy Storage (CAES) also works as a generation storage technology by using the elastic potential energy of compressed air to improve the efficiencies of conventional gas turbines.

CAES systems compress air using electricity during off-peak times, and then store the air in underground caverns. During times of peak demand, the air is drawn from storage and fired with natural gas in a combustion turbine to generate electricity. This method uses only a third of the natural gas used in conventional methods. Because CAES plants require some sort of underground reservoir, they are limited by their locations. Two commercial CAES plants currently operate in Huntorf, Germany and MacIntosh, Alabama, though plants have been proposed in other parts of the United States.

### Stored Hydrogen

Hydrogen can be used as a zero-carbon fuel for generation. Excess electricity can be used to create hydrogen, which can be stored and used later in fuel cells, engines, or gas turbines to generate electricity without producing harmful emissions. NREL has studied the potential or creating hydrogen from wind power and storing it in the wind turbine towers for electricity generation when the wind isn't blowing.

### Pumped Hydroelectric Storage

Pumped hydroelectric storage offers a way to store energy at the grid's transmission stage, by storing excess generation for later use.

Many hydroelectric power plants include two reservoirs at different elevations. These plants store energy by pumping water into the upper reservoir when supply exceeds demand. When demand exceeds supply, the water is released into the lower reservoir by running downhill through turbines to generate electricity.

With more than 22 GW of installed capacity in the United States, pumped hydro storage is the largest storage system operating today. However, the long permitting process and high cost of pumped storage makes further projects unlikely.

### Flywheels

Flywheels can provide a variety of benefits to the grid at either the transmission or distribution level, by storing electricity in the form of a spinning mass.

The device is shaped liked a cylinder and contains a large rotor inside a vacuum. When the flywheel draws power from the grid, the rotor accelerates to very high speeds, storing the electricity as rotational energy. To discharge the stored energy, the rotor switches to generation mode, slows down, and runs on inertial energy, thus returning electricity to the grid.

Flywheels typically have long lifetimes and require little maintenance. The devices also have high efficiencies and rapid response times. Because they can be placed almost anywhere, flywheels can be located close to the consumers and store electricity for distribution.

While a single flywheel device has a typical capacity on the order of kilowatts, many flywheels can be connected in a "flywheel farm" to create a storage facility on the order of megawatts. Beacon Power's Stephentown Flywheel Energy Storage Plant in New York is the largest flywheel facility in the United States, with an operating capacity of 20 MW.

### Batteries

Like flywheels, batteries can be located anywhere so they are often seen as storage for distribution, when a battery facility is located near consumers to provide power stability; or end-use, such as batteries in electric vehicles.

There are many different types of batteries that have large-scale energy storage potential, including sodium-sulfur, metal air, lithium ion, and lead-acid batteries. Over 80% of U.S. large-scale battery storage power capacity is currently provided by batteries based on lithium-ion chemistries. About 90% of large-scale battery storage in the United States is installed in regions covered by five of the seven organized independent system operators (ISOs) or regional transmission organizations (RTOs) and in Alaska and Hawaii (AK/HI). There are several battery installations at wind farms; including the Notrees Wind Storage Demonstration Project in Texas, which uses a 36 MW battery facility to help ensure stability of the power supply even when the wind isn't blowing. At the end of 2017, 708 megawatts (MW) of power capacity, representing 867 megawatt hours (MWh) of energy capacity, of large-scale battery storage capacity was in operation. (EIA, 2019)

### HARDENING THE GRID

The nation's power grid is vulnerable to the effects of an electromagnetic pulse (EMP), a sudden burst of electromagnetic radiation resulting from a natural or man-made event. EMP events occur with little or no warning and can have catastrophic effects, including causing outages to major portions of the U.S. power grid possibly lasting for months or longer. Naturally occurring EMPs are produced as part of the normal cyclical activity of the sun while man-made EMPs, including Intentional Electromagnetic Interference (IEMI) devices and High Altitude Electromagnetic Pulse (HEMP), are produced by devices designed specifically to disrupt or destroy electronic equipment or by the detonation of a nuclear device high above the earth's atmosphere. EMP threats have the potential to cause wide scale long-term losses with economic costs to the United States that vary with the magnitude of the event.

Naturally occurring EMP events resulting from magnetic storms that flare on the surface of the sun are inevitable. Although we do not know when the next significant solar event will occur, we do know that the geomagnetic storms they produce have occurred at varying intensities throughout history. We are currently entering an interval of increased solar activity and are likely to encounter an increasing number of geomagnetic events on earth. (Electromagnetic Pulse: Effects on the U.S. Power Grid FERC executive summary, 2019)

## OFF-GRID

Energy storage also becomes more important the farther you are from the electrical grid. For example, when you turn on the lights in your home, the power comes from the grid; but when you turn on a flashlight while camping, you must rely on the stored energy in the batteries. Similarly, homes that are farther away from the transmission grid are more vulnerable to disruption than homes in large metropolitan areas. Islands and microgrids that are disconnected from the larger electrical grid system depend on energy storage to ensure power stability, just like you depend on the batteries in your flashlight while camping.

The primary Off-Grid energy requirement is for transportation. Carbon-based fuels are the only viable energy source for aircraft, and marine(non-military), and most commercial and personal vehicle use. This includes construction and agriculture equipment. Government intervention in personal vehicle use by way of subsidies, grants and loan guarantees has artificially accelerated the use of electric vehicles. However, it is unlikely than the current renewable energy sources will be able to meet these needs until a light-weight, efficient storage system with low environmental impact is developed.

## MILITARY AND DEFENSE REQUIREMENTS

Military and national defense requirements are almost entirely Off-Grid for obvious reasons. Civilian networks are vulnerable to outside forces and most military logistics and operations require extreme portability and reliability. There is an advantage to be gained in the funding of research directed toward military and defense energy needs. The disadvantage is the timeframe for release of the advances in technology to reach the civilian and private marketplace. This is an acceptable disadvantage since the proposed transition away from carbon-based fuels will not be on the same timeline as the overall U. S. Energy plan. Therefore, this topic will be excluded from further discussion in this book. In my opinion, military and defense involvement in the U. S. Energy plan will need to be one of cooperative and efficient liaison.

# THE PLANNING PROCESS

## THE DEPARTMENT OF ENERGY

The United States Department of Energy (DOE) is a cabinet-level department of the United States Government concerned with the United States' policies regarding energy and safety in handling nuclear material. Its responsibilities include the nation's nuclear weapons program, nuclear reactor production for the United States Navy, energy conservation, energy-related research, radioactive waste disposal, and domestic energy production. It also directs research in genomics; the Human Genome Project originated in a DOE initiative. DOE sponsors more research in the physical sciences than any other U.S. federal agency, the majority of which is conducted through its system of National Laboratories. The agency is administered by the United States Secretary of Energy, and its headquarters are located in Southwest Washington, D.C., on Independence Avenue in the James V. Forrestal Building, named for James Forrestal, as well as in Germantown, Maryland.

### *HISTORY*

In 1942, during World War II, the United States started the Manhattan Project, a project to develop the atomic bomb, under the eye of the U.S. Army Corps of Engineers. After the war in 1946, the Atomic Energy Commission (AEC) was created to control the future of the project. Among other nuclear projects, the AEC produced fabricated uranium fuel cores at locations such as Fernald Feed Materials Production Center in Cincinnati, Ohio. In 1974, the AEC gave way to the Nuclear Regulatory Commission, which was tasked with regulating the nuclear power industry, and the Energy Research and Development

Administration, which was tasked to manage the nuclear weapon, naval reactor, and energy development programs.

The 1973 oil crisis called attention to the need to consolidate energy policy. On August 4, 1977, President Jimmy Carter signed into law The Department of Energy Organization Act of 1977, which created the Department of Energy. The new agency, which began operations on October 1, 1977, consolidated the Federal Energy Administration, the Energy Research and Development Administration, the Federal Power Commission, and programs of various other agencies. Former Secretary of Defense James Schlesinger, who served under Presidents Nixon and Ford during the Vietnam War, was appointed as the first secretary. (Wkipedia, 2019)

### THE PRESENT

Today, the Department of Energy (DOE) has 11 Program offices, 19 Staff offices, 21 Laboratories and Technology centers, 4 Regional Power Administration offices, 10 Field sites, and two other Agencies, The Energy Information Administration and the National Nuclear Security Administration. This large organization has an incredibly mixed mission statement. The organizational structure is not conducive to the development of an overall National Energy plan since there is extreme redundancy in the Program and Staff offices.

There have been excellent changes under the new Administration. Modeled after the Defense Advanced Research Projects Agency (DARPA), the Energy Department's new Advanced Research Projects Agency-Energy (ARPA-E) funds game-changing energy technologies that are typically too early for private-sector investment.

From new wind turbine designs and transportation fuels made from bacteria to innovative energy storage solutions and smaller, more efficient semiconductors, ARPA-E projects have the potential to change the way we generate, store and use energy. The program

105

continues to invest in technologies that could radically improve U.S. economic prosperity, national security and environmental well-being. (Department of Energy ARPA-E)

The new Advanced Research Projects Agency-Energy organization has been busy. Last December (2018) ARPA-E announced $21 million in funding for 7 projects in the next two cohorts of the agency's OPEN+ program. These project teams will pursue methods to create high-value carbon and hydrogen from methane, or to produce super strong, durable concrete with lower cost and environmental impact. It also announced an $18 Million funding opportunity designed to support early stage, transformative energy technologies. The "Solicitation on Topics Informing New Program Areas" funding opportunity enables ARPA-E to investigate potential new program areas while highlighting energy challenges of critical interest to American competitiveness and security. On January 15, 2019 ARPA-E announced $11 million in funding for 7 projects in the next two cohorts of the agency's OPEN+ program: Energy-Water Technologies and Sensors for Bioenergy and Agriculture. (ARPA-E , 2019)

However, is unlikely that the current organizational structure, competition for human and financial resources (funding) and partisan political environment will allow the DOE to develop a long-range plan for the transition from carbon-based energy source and infrastructure to a different energy source and a suitably modified energy infrastructure. Therefore, the organization must be realigned in a logical manner in order for it to fulfill its duty to the American public. It must have a goal. If one of its goals is to manage the transition of the country away from carbon-based fuels i.e. fossil fuels, then it should be structured to do just that.

The first element of an Energy Plan must include reasonable benchmarks for each main non-renewable energy supply component.

The supplier of the energy source must reduce pollution incurred in the resource production process. The power generator must reduce pollution created in the electrical (or perhaps) generation process. The ultimate end user must use the secondary energy source (electricity or hydrogen) efficiently and reduce pollution from its use.

The process to transition away from carbon-based fuels should be considered a major project. This will involve determining the activities, durations, and costs to get to the final objective. A significant complication in this project is the actual unknown makeup of the ultimate energy source. Scientific breakthroughs that will likely occur during the project timetable will require major revisions to the Plan and the ultimate energy destination may be completely different than currently anticipated.

The role of government should be primarily one of coordination between the supplier, generator and ultimate end-user of the energy source. The benchmarks in the plan should be agreed to in a democratic process with input from various experts provided by the vested interests of the suppliers, generators and end-users. Government intervention should be avoided at all costs since this often deteriorates to bureaucrats defining "winners and losers" for political reasons. There will be Government end-users in the mix such as the military establishment. That should be sufficient. Political ideologies must be excluded from the planning process at all cost.

The optimum role for existing government bureaucracies involved with energy should be to provide information resources and support research outlined by the U. S. Energy Plan developers and those involved in its execution. The mission of each existing bureaucracy should be reviewed and revised to fit under the umbrella of the Energy Plan. The head of the U. S. Energy Plan Administration should be a non-political position who reports directly to the Secretary of Energy. The Senate and House of Representatives should set up specific subcommittees for oversight of the Energy Plan and should be designated as the U. S. Energy Plan oversight subcommittees only. The majority leaders should select the chairs and minority leaders of the subcommittees from their best technical members with proven non-partisan records in developing legislation.

## MERGING THE EIA WITH ARPA-E TO FORM THE NEW
## U. S. ENERGY PLANNING ADMINISTRATION

With those guidelines in place, the first organizational step will be to rename one existing agency of the DOE, the U. S. Energy Information Administration to the U. S. Energy Planning Administration (USEPA). The Advanced Research Projects Agency-Energy organization (ARPA-E) should then be merged with the new USEPA. While this may sound simplistic on the surface, the current EIA has the information resources and industry contacts that would facilitate the development of an Energy Plan, it just does not have that as its mission. ARPA-E is already providing funding for advancing new technologies. Most of the reference material presented in this book was downloaded from the EIA website which reinforces the point that this Agency has at hand the tools to develop a logical, cohesive plan to transition away from carbon-based fuels to an alternative energy source. Merging EIA and ARPA-E into a new U. S. Energy Planning Administration is an obvious fit and will reduce redundancy and improve efficiency.

Once the new U.S. Energy Planning Administration is stood up, the respective Senate and Congressional Energy Plan oversight subcommittees should second skilled staff to the new Administration to begin drafting legislation to support the plan development. To get things underway, the draft proposal for the Green New Deal should be rejected and replaced with something along the lines of the proposed legislation that follows in the next chapter of this book.

The Secretary of the Department of Energy is an executive appointed by the President of United States and confirmed by the Senate. This executive should be able to realign the organization without political interference. However, the Chief Executive may want to initiate the realignment first so that the House and Senate are stimulated to start drafting supporting legislation.

This author does not pretend to be qualified to write legislation. But, apparently, neither are the drafters of the Green New Deal. However, the guidelines for properly preparing draft legislation are available at the Office of the Legislative Counsel's website at http://legcounsel.house.gov/HOLC/Resources/comps_alpha.html.

The following chapter is an initial attempt using those guidelines. (Quick Guide to Legislative Drafting Office of the Legislative Counsel, U.S. House of Representatives, 2019) This is presented  as a less onerous way to implement a program to transition to a new energy source than that proposed in the draft Green New Deal legislation.

# ENERGY PLAN ADDENDUM TO HOUSE RULES

DRAFT TEXT FOR PROPOSED ADDENDUM TO HOUSE RULES FOR 116TH CONGRESS OF THE UNITED STATES SEC. [_____]. COMMITTEES, COMMISSIONS, AND HOUSE OFFICES. (a) Establishment of the select oversight subcommittee for the new U. S. Energy Planning Administration (USEPA).

(1) ESTABLISHMENT; COMPOSITION.—(A) ESTABLISHMENT.—There is hereby established a select subcommittee (hereinafter in this section referred to as the "select subcommittee"). (B) COMPOSITION.—The select subcommittee shall be composed of 5 members appointed by the Speaker, of whom 2 may be appointed on the recommendation of the Minority Leader. The Speaker shall designate one member of the select subcommittee as its chair. A vacancy in the membership of the select subcommittee shall be filled in the same manner as the original appointment.

(2) JURISDICTION; FUNCTIONS.— (A) LEGISLATIVE JURISDICTION.— The select subcommittee shall have oversight of the work by the new U. S. Energy Planning Administration (USEPA) in the reorganization of U. S. Department of Energy to facilitate the development of a detailed national, industrial, economic mobilization plan (hereinafter in this section referred to as the "U. S. Energy Plan" or the "Plan") for the reasonable transition of the United States energy base from carbon-based fuels to a reliable, perhaps renewable energy source yet to be determined. In furtherance of the foregoing, the subcommittee will mirror the oversight performed by the Senate select

subcommittee of the same name. In addition to oversight, the select committee shall provide or second skilled support staff to the new U. S. Energy Planning Administration to draft legislation for the enactment of elements of the Energy Plan that require government resource allocations and funding that would be in compliance with the overall U. S. Budget and Deficit Reduction Plans. The select subcommittee shall not have legislative jurisdiction and shall have no authority to take legislative action on any bill or resolution. (B) INVESTIGATIVE JURISDICTION.— The select subcommittee shall have no authority to investigate.

(3) PROCEDURE.— (A) The select subcommittee shall have the authorities and responsibilities of, and shall be subject to the same limitations and restrictions as, a standing subcommittee of the House, and shall be deemed a subcommittee of the House for all purposes of law or rule. (B)(i) Rules [to be confirmed by reference to overall House Rules package] (Organization of Committees) and [to be confirmed by reference to overall House Rules package] (Procedures of Committees and Unfinished Business) shall apply to the select subcommittee where not inconsistent with this resolution. (ii) Service on the select subcommittee shall not count against the limitations on committee or subcommittee service in Rule [to be confirmed by reference to overall House Rules package] (Organization of Committees).

(4) FUNDING.—To enable the select subcommittee to carry out the purposes of this section— (A) The select subcommittee may use the services of staff of the House and may, with the approval of the Speaker of the House use the services of external consultants or experts in furtherance of its mandate; (B) The select subcommittee shall not be eligible for interim funding. (C) The subcommittee will apply to the

House of Representatives for a dedicated budget to carry out its mandate.

(5) INTERIM REPORTING; PROGRESS BY U. S. ENERGY PLANNING ADMINISTRATION, SUBMISSION OF DRAFT LEGISLATION— (A) The select subcommittee may periodically report the results of its studies to the House or any House Committee it deems appropriate.

(B) (i.) The select committee shall work with the head of the U. S. Energy Planning Administration (USEPA) to complete a draft Organizational and Administrative Plan (OAP) for all existing U. S. Departments involved with Energy to transfer all Energy planning personnel to USEPA by a date no later than January 1, 2020. (ii.) The select subcommittee shall complete its review and approve the finalized draft OAP by a date no later than the date that is 90 calendar days after the USEPA has completed the OAP and, in any event, no later than March 1, 2020. (iii) The select subcommittee shall ensure that the draft OAP is made available to the Executive branch for review and suggestions for revision as appropriate. The draft OAP will not be made available to the general public unless authorized by the Executive branch.

(6) SCOPE OF THE ORGANIZATIONAL AND ADMINISTRATIVE PLAN FOR THE U. S. ENERGY PLANNING ADMINISTRATION (A) The ORGANIZATIONAL AND ADMINISTRATIVE PLAN FOR THE U. S. ENERGY PLANNING ADMINISTRATION (USEPA) (and the draft legislation) shall be developed with the objective of providing the appropriate resources and establishing reporting relationships for all U. S. departments involved with Energy.

(i.) Working with the interim head of the new U. S. Energy Planning Administration USEPA, the subcommittee will identify duplicate planning activities in other departments and propose the consolidation of those activities under USEPA. It is anticipated that the consolidations will involve the transfer and relocation of personnel with specialized skills.

(ii.) The subcommittee will request and receive periodic updates on the progress of both administration of USEPA and the development of the Energy Plan.

(iii.) The subcommittee will request and receive an operating budget from the head of the USEPA that will be developed from a zero baseline each fiscal year.

(iv). The subcommittee will prepare its own status report to facilitate briefing the Congress on the Energy Plan progress of the USEPA.

(B) The subcommittee will coordinate with its counterpart in the U. S. Senate on a quarterly basis. It will provide briefings to the Executive branch semi-annually. Working with the USEPA it will:

(i.) ensure that the Energy Plan provides all members of our society, across all regions and all communities, the opportunity, training and education to be a full and equal participant in the transition away from carbon-based fuels.

(ii.) Identify communities where the transition from carbon-based fuels may require economic assistance to those communities...

# Get the Idea?

> *Any politician of any stripe who buys into the GREEN NEW DEAL and the implementation of a Carbon Tax has not performed "due diligence" and should not be re-elected.*

# ABOUT THE AUTHOR

Martin Capages, Jr. is a retired professional engineer, technical executive and an Army veteran. His technical and management experience includes aircraft design, petroleum exploration and production, computer modeling and technology applications and structural engineering. He began writing political commentary in 2009 and completed his first book, *The Moral Case for American Freedom*, in July 2017. His writing is from the perspective of an engineer, Christian layman, conservative and Constitutional originalist.

Martin attended Missouri State University and the Missouri University of Science and Technology where he graduated with a Bachelor of Science in Mechanical Engineering in 1967. After receiving his Commission as an Army Ordnance Officer but prior to reporting for active duty, he joined Boeing Aircraft in Wichita as an Associate Engineer working on the new 737. He reported for active duty in June 1967. After completing active duty, Martin joined Exxon in Houston, Texas, with assignments throughout the U.S. and Europe to include serving as acting North Sea Development Planning Manager for Exxon in London, Production Operations in the Gulf of Mexico, Engineering Manager for the Texas Midland District, the Alaska Financial and Facilities groups, and Exxon's Western Division Computing organization. He left Exxon in 1984 to join Kerr McGee in Oklahoma as Manager of Engineering Services until 1992 when he left the petroleum industry to start his own structural engineering consulting firm, ARIS Engineering Inc., in Springfield, Missouri. He continued post-graduate studies in Civil Engineering and Management receiving an earned Doctorate in Engineering Management in 2002. He retired from full time practice in 2012.

Martin is married to Pamela Kay Capages.  They have five children and seven grandchildren. Both Martin and Pamela are active members of their local Baptist church and serve in other state and international Christian ministries. Pamela is an author in her own right and has published books of poetry concerning her Christian faith, family and personal observations of nature.

# WORKS CITED

*AOC Green New Deal.* (2019, January 10). Retrieved from www.nakedcapitalism: https://www.nakedcapitalism.com/wp-content/uploads/2019/01/AOC_Green_New_Deal.pdf

*ARPA-E* . (2019, January 10). Retrieved from energy.gov: https://www.energy.gov/science-innovation/innovation/arpa-e

Bell, L. (2015, February 5). *In their own words climate alarmists debunk their science.* Retrieved from forbes.com: https://www.forbes.com/sites/larrybell/2013/02/05/in-their-own-words-climate-alarmists-debunk-their-science/#7a83b99568a3

Cherry, R. R. (2018). *Restoring the American Mind.* Springfield, MIssouri: American Freedom Publications LLC.

Department of Energy ARPA-E. (n.d.). Retrieved from https://arpa-e.energy.gov/

Deutch, J. (2019, January 28). *Demonstrating Near Carbon-Free Electricity Generation from Renewables and Storage.* Retrieved from Stanford.edu: https://earth.stanford.edu/.../energy-seminar-john-deutch-demonstrating-near- carbon-free-electricity-generation-renewables

EIA. (2019, January 7). *Energy Explained.* Retrieved from eia.gov: https://www.eia.gov/energyexplained/index.php?page=about_home

EIA. (2019, January 16). *Energy Outlooks.* Retrieved from eia.gpv: https://www.eia.gov/outlooks/aeo/data/browser/#/?id=1-AEO2018&cases=ref2018&sourcekey=0

EIA. (2019, January 10). *Today in Energy.* Retrieved from eia.gov: https://www.eia.gov/todayinenergy/detail.php?id=37034#

*EIA on Nuclear Fuel.* (2019, January 17). Retrieved from https://www.eia.gov/energyexplained: https://www.eia.gov/energyexplained/index.php?page=nuclear_fuel_cycle

*EIA on Nuclear Power Plants.* (2019, January 16). Retrieved from www.eia.gov/energyexplained: https://www.eia.gov/energyexplained/index.php?page=nuclear_power_plants

*Electromagnetic Pulse: Effects on the U.S. Power Grid ferc executive summary.* (2019, January 16). Retrieved from https://www.ferc.gov: https://www.ferc.gov/industries/electric/indus-act/reliability/cybersecurity/ferc_executive_summary.pdf

Epstein, A. (2014). *The Moral Case for Fossil Fuels.* New York: Penguin Group.

Epstein, A. (2019, January 12). *Why Green Energy Means No Energy.* Retrieved from industrialprogress.com: http://industrialprogress.com/why-green-energy-means-no-energy/

Fyfe, J. C., Meehl, G. A., England, M. H., Mann, M. E., Santer, B. D., Flato, G. M., . . . Swart, N. C. (2016, March). Making sense of the early 2000s warming slowdown. *Nature Climate Change, 6*, 224-228. Retrieved from http://www.meteo.psu.edu/holocene/public_html/Mann/articles/articl es/FyfeEtAlNatureClimate16.pdf

Lundseng, O., Johnsen, H., & Bergsmark, S. (2019, January 17). *Germany's Green Transition has hit a brick wall.* Retrieved from wattsupwiththat.com: https://wattsupwiththat.com/2018/12/21/germanys-green-transition-has-hit-a-brick-wall/

May, A. (2018). *CLIMATE CATASTROPHE: Science or Science Fiction?* The Woodlands Texas: American Freedom Publications LLC.

Metcalf, T. (2017, December 29). *Long wait for fusion may be coming to an end.* Retrieved from nbcnews.com: https://www.nbcnews.com/mach/science/long-wait-fusion-power-may-be-coming-end-ncna833251

Newell, R., & Dopplick, T. (1979). Questions Concerning the Possible Influence of Anthropogenic CO2 on Atmospheric Temperature. *J. Applied Meterology, 18*, 822-825. Retrieved from http://journals.ametsoc.org/doi/pdf/10.1175/1520-0450(1979)018%3C0822%3AQCTPIO%3E2.0.CO%3B2

*Pathological Science.* (2019, January 14). Retrieved from wikipedia.org: https://en.wikipedia.org/wiki/Pathological_science

*Quick Guide to Legislative Drafting Office of the Legislative Counsel, U.S. House of Representatives.* (2019, January 17). Retrieved from legcounsel.house.gov.: http://legcounsel.house.gov/HOLC/Resources/comps_alpha.html.

Rep. Gabbard, Tulsi [D-HI-2]. (2017, September 1). *OFF Act .* Retrieved from congress.gov: https://www.congress.gov/bill/115th-congress/house-bill/3671

Scotese, C. (2015). *Some thoughts on Global Climate Change: The Transition from Icehouse to Hothouse.* PALEOMAP Project. Retrieved from https://www.researchgate.net/profile/Christopher_Scotese3/project/E arth-History-The-Evolution-of-the-Earth-System/attachment/575023e708aec90a33750af1/AS:36850507034214 4@1464869863189/download/Some+Thoughts+on+Global+Climate+Ch angev21ar+copy.pdf

Singh, S. (2015, September 23). *How long will fossil fuel last.* Retrieved from business standard: https://www.business-standard.com/article/punditry/how-long-will-fossil-fuels-last-115092201397_1.html

Sullivan, M. (2004, August 10). *FDR's policies prolonged Depression by 7 Years.* Retrieved from newsroom.ucla.edu: http://newsroom.ucla.edu/releases/FDR-s-Policies-Prolonged-Depression-5409

*U. S. Coal Reserves.* (2019, January 10). Retrieved from eia.gov: https://www.eia.gov/energyexplained/index.php?page=coal_reserves

*U. S. Energy Information Administration( EIA) Home Page.* (2019, January 16). Retrieved from https://www.eia.gov/energyexplained/index: https://www.eia.gov/energyexplained/index

Union of Concerned Scientists (US). (2019, January 15). *How Energy Storage Works.* Retrieved from ucsusa.org: https://www.ucsusa.org/clean-energy/how-energy-storage-works#bf-toc-2

Wikipedia. (2019, January 10). *The Green New Deal.* Retrieved from wikipedia.org: https://en.wikipedia.org/wiki/Green_New_Deal

Wkipedia. (2019, January 14). *Biomass.* Retrieved from wikipedia.org: https://en.wikipedia.org/wiki/Biomass

Wkipedia. (2019, January 12). *U. S. Department of Energy .* Retrieved from wikipedia.org:
https://en.wikipedia.org/wiki/United_States_Department_of_Energy

Wyden, S. R. (2019, January 10). *It's Time for a 'Green New Deal' .* Retrieved from Politico.com:
https://www.politico.com/magazine/story/2019/01/10/green-new-deal-congress-ron-wyden-223910

Since the American principle of rightful human liberty derived from equal human rights is self-evident moral truth, and the selective superior rights and associated infringement of human liberty in socialism is self-evidently wrong, the question arises: How are equal rights and rightful human liberty protected? Thomas Jefferson answers: Equality before law protects equal human rights and therefore secures rightful human liberty. (Cherry, 2018)

--- *Restoring the American Mind*, Ronald R. Cherry MD 2018

"Bear in mind this sacred principle, that though the will of the majority is in all cases to prevail, that will, to be rightful, must be reasonable; that the minority possess their equal rights [along with the majority], which equal laws must protect, and to violate would be oppression." Thomas Jefferson

# INDEX

CPSIA information can be obtained
at www.ICGtesting.com
Printed in the USA
LVHW010915180219
607863LV00002B/157/P